Springer Undergraduate Mathematics Series

The Springer Undergraduate Mathematics Series (SUMS) is a series designed for undergraduates in mathematics and the sciences worldwide. From core foundational material to final year topics, SUMS books take a fresh and modern approach. Textual explanations are supported by a wealth of examples, problems and fully-worked solutions, with particular attention paid to universal areas of difficulty. These practical and concise texts are designed for a one- or two-semester course but the self-study approach makes them ideal for independent use.

M. A. Al-Gwaiz

Sturm–Liouville Theory and its Applications

Second Edition

 Springer

M. A. Al-Gwaiz
Department of Mathematics
King Saud University
Riyadh, Saudi Arabia

ISSN 1615-2085   ISSN 2197-4144   (electronic)
Springer Undergraduate Mathematics Series
ISBN 978-1-4471-7609-1   ISBN 978-1-4471-7610-7   (eBook)
https://doi.org/10.1007/978-1-4471-7610-7

Mathematics Subject Classification: 26-01, 34-01

New edition of "Sturm-Liouville Theory and its Applications", Springer, 2008

Preface to the Second Edition

I started thinking about revising the first edition of this book early last year, 16 years after it was published, and almost 10 years after I had retired from teaching at King Saud University. Over the years I had received a number of comments and queries from former students and others whom I did not know, who seemed to have taken a keen interest in the subject. To me the book has always been very special. I was deeply involved in designing and developing the course on which it was based, and which I taught for many years to young third and fourth year undergraduates. In the 1990s, I used an Arabic version of this book entitled *Mathematical Methods of Fourier Analysis*, based on my lecture notes at the time.

An undergraduate textbook on the Sturm-Liouville theory and Fourier analysis, in my opinion, needs to balance readability and motivation with the rigours of abstract analysis. That has always been my conviction, and it remains so today. Now, it seemed, was an opportunity to revisit the subject. This edition differs from the first in two main respects. The first is the addition of a brief outline of Lebesgue measure and integration as an Appendix at the end of the book, including some theorems and other results that have relevance to the subject matter, such as the dominated convergence theorem and the Riemann-Lebesgue lemma. This was done mainly to place the completion of the space $\mathcal{L}^2$ on a more solid basis, and to reduce the burden of proof in a few instances by using these well-established results of real analysis. The second is the opportunity to rearrange some topics, correct mistakes and misprints that I had noticed, or had been pointed out to me, and to fine-tune the general presentation, which a revised edition always offers.

I should like to extend my gratitude to Gordon McLean for sharing his thoughts and suggestions on the first edition, and to my old colleague from the bygone days of KSU, Professor Saleh Elsanousi, for some useful exchanges that we had on the new draft. Mohammed Balfageh was most helpful in retrieving the files of the first edition from my old laptop, which was no easy task, and in preparing the LaTeX file for the new edition. My son Majid checked the solutions to many of the exercises, and the rest of the family was equally understanding and supportive during the last few months of my preoccupation with this project. Finally, to my contacts at Springer, I wish to extend my sincere appreciation and respects for their care and attention.

Riyadh, Saudi Arabia   M. A. Al-Gwaiz
June 2025

Preface to the First Edition

This book is based on lecture notes which I have used over a number of years to teach a course on Mathematical Methods to senior undergraduate students of mathematics at King Saud University. The course is offered here as a prerequisite for taking partial differential equations in the final (fourth) year of the undergraduate program. It was initially designed to cover three main topics: Special functions, Fourier series and integrals, and a brief sketch of the Sturm-Liouville problem and its solutions. Using separation of variables to solve a boundary-value problem for a linear second order partial differential equation often leads to a Sturm-Liouville eigenvalue problem, and the solution set is likely to be a sequence of special functions, hence the relevance of these topics. Typically, the solution of the partial differential equation can then be represented (pointwise) by a Fourier series or a Fourier integral, depending on whether the domain is finite or infinite.

But it soon became clear that these "mathematical methods" could be developed into a more coherent and substantial course by presenting them within the more general Sturm-Liouville theory in $\mathcal{L}^2$. According to this theory, a linear second order differential operator which is self-adjoint has an orthogonal sequence of eigenfunctions which spans $\mathcal{L}^2$. This immediately leads to the fundamental theorem of Fourier series in $\mathcal{L}^2$ as a special case in which the operator is simply d^2/dx^2. The other orthogonal functions of mathematical physics, such as the Legendre and Hermite polynomials or the Bessel functions, are similarly generated as eigenfunctions of particular differential operators. The result is a generalized version of the classical theory of Fourier series, which ties up the topics of the course mentioned above and provides a common theme for the book.

In Chap. 1 the stage is set by defining the inner product space of square integrable functions $\mathcal{L}^2$, and the basic analytical tools needed in the chapters to follow. These include the convergence properties of sequences and series of functions and the important notion of completeness of $\mathcal{L}^2$, which is defined through Cauchy sequences.

The difficulty with building Fourier analysis on the Sturm-Liouville theory is that the latter is deeply rooted in functional analysis, in particular the spectral theory of compact operators, which is beyond the scope of an undergraduate treatment such as this. We need a simpler proof of the existence and completeness of the eigenfunctions. In the case of the regular Sturm-Liouville problem, this is achieved in Chap. 2 by invoking the existence theorem for linear differential equations to construct Green's function for the Sturm-Liouville operator, and then using the Ascoli-Arzela theorem to arrive at the desired conclusions. This is covered in Sects. 2.4.1 and 2.4.2 along the lines of Coddington and Levinson in [5].

Chapters 3, 4, and 5 present special applications of the Sturm-Liouville theory. Chapter 3, which is on Fourier series, provides the prime example of a regular Sturm-Liouville problem. In this chapter the pointwise theory of Fourier series is also covered, and the classical theorem (Theorem 3.9) in this context is proved. The advantage of the $\mathcal{L}^2$ theory is already evident from the simple statement of Theorem 3.2, that a function can be represented by a Fourier series if and only if it lies in $\mathcal{L}^2$, as compared to the statement of Theorem 3.9.

In Chaps. 4 and 5 we discuss some of the more important examples of a singular Sturm-Liouville problem. These lead to the orthogonal polynomials and Bessel functions which are familiar to students of science and engineering. Each chapter concludes with applications to some well known equations of mathematical physics, including Laplace's equation, the heat equation, and the wave equation.

Chapters 6 and 7 on the Fourier and Laplace transformations are not really part of the Sturm-Liouville theory, but are included here as extensions of the Fourier series method for representing functions. These have important applications in heat transfer and signal transmission. They also allow us to solve non-homogeneous differential equations, a subject which is not discussed in the previous chapters where the emphasis is mainly on the eigenfunctions.

The reader is assumed to be familiar with the convergence properties of sequences and series of functions, which are usually presented in advanced calculus, and with elementary ordinary differential equations. In addition, we have used some standard results of real analysis, such as the density of continuous functions in $\mathcal{L}^2$ and the Ascoli-Arzela theorem. These are used to prove the existence of eigenfunctions for the Sturm-Liouville operator in Chap. 2, and they have the advantage of avoiding any need for Lebesgue measure and integration. It is for that reason that smoothness conditions are imposed on the coefficients of the Sturm-Liouville operator, for otherwise integrability conditions would have sufficed. The only exception is the dominated convergence theorem, which is invoked in Chap. 6 to establish the continuity of the Fourier transform. This is a marginal result which lies outside the context of the Sturm-Liouville theory and could have been handled differently, but the temptation to use that powerful theorem as a shortcut was irresistible.

This book follows a strict mathematical style of presentation, but the subject is important for students of science and engineering. In these disciplines, Fourier analysis and special functions are used quite extensively for solving linear differential equations, but it is only through the Sturm-Liouville theory in $\mathcal{L}^2$ that one discovers the underlying principles which clarify why the procedure works. The theoretical treatment in Chap. 2 need not hinder students outside mathematics who may have some difficulty with the analysis. Proof of the existence and completeness of the eigenfunctions (Sects. 2.4.1 and 2.4.2) may be skipped by those who are mainly interested in the results of the theory. But the operator-theoretic approach to differential equations in Hilbert space has proved extremely convenient and fruitful in quantum mechanics, where it is introduced at the undergraduate level, and it should not be avoided where it seems to bring clarity and coherence in other disciplines.

I have occasionally used the symbols $\Rightarrow$ (for "implies") and $\Leftrightarrow$ (for "if and only if") to connect mathematical statements. This is done mainly for the sake of typographical convenience and economy of expression, especially where displayed relations are involved.

A first draft of this book was written in the summer of 2005 while I was on vacation in Lebanon. I should like to thank the librarian of the American University of Beirut for allowing me to use the facilities of their library during my stay there. A number of colleagues in our department were kind enough to check the manuscript for errors and misprints, and to comment on parts of it. I am grateful to them all. Professor Saleh Elsanousi prepared the figures for the book, and my former student Mohammed Balfageh helped me to set up the software used in the SUMS Springer series. I would not have been able to complete these tasks without their help. Finally, I wish to express my deep appreciation to Karen Borthwick at Springer-Verlag for her gracious handling of all the communications leading to publication.

Riyadh, Saudi Arabia   M. A. Al-Gwaiz
March 2007

Declarations

Competing Interests The authors have no competing interests to declare that are relevant to the content of this manuscript.

Contents

Notation

$\mathbb{N} = \{1, 2, 3, \cdots\}$
$\mathbb{N}_0 = \{0, 1, 2, 3, \cdots\}$
$\mathbb{Z} = \{\cdots, -2, -1, 0, 1, 2, \cdots\}$

$\mathbb{Q}$	rational numbers
$\mathbb{R}$	real numbers
$\mathbb{C}$	complex numbers
$\mathbb{F}$	$\mathbb{R}$ or $\mathbb{C}$
$C(I)$, $C^0(I)$	set of continuous functions on the interval I
$C^m(I)$	functions on I which have continuous derivatives up to order m
$C^\infty(I)$	functions on I which have continuous derivatives of all orders
$\mathcal{P}(I)$, $\mathcal{P}_n(I)$	polynomials on I, polynomials on I of degree n
$\mathcal{E}$	locally integrable functions on $[0, \infty)$ of exponential growth
m^*	outer measure
m	Lebesgue measure
$\mathcal{M}$	Lebesgue measurable sets in $\mathbb{R}$
χ_E	characteristic function of E
$\mathcal{S}, \mathcal{S}_+$	Simple functions, positive simple functions
$\mathcal{L}^1, \mathcal{R}$	Lebesgue integrable functions, Riemann integrable functions
$\mathcal{L}^2, \mathcal{L}^2_\rho$	Lebesgue square-integrable functions, and with weight ρ
$\langle \cdot, \cdot \rangle$, $\langle \cdot, \cdot \rangle_\rho$	inner product in $\mathcal{L}^2$, and with weight ρ
$\|\cdot\|$, $\|\cdot\|_\rho$	$\mathcal{L}^2$ norm
$f_n \to f$	the sequence of functions f_n converges pointwise to f
$f_n \xrightarrow{u} f$	f_n converges uniformly to f
$f_n \xrightarrow{\mathcal{L}^2} f$	f_n converges to f in $\mathcal{L}^2$
$\hat{f} = \mathcal{F}(f)$	Fourier transform of f
$F = \mathcal{L}(f)$	Laplace transform of f
$f * g$	convolution of f and g
$f^\pm$	positive and negative parts of the function f
L'	adjoint of the differential operator L
L^*	formal adjoint of L

Δ	Laplacian operator
$W(f, g)$	Wronskian of f and g
$G(x, \xi)$	Green's function
$D_n(\alpha)$	Dirichlet's kernel
erf	error function
P_n	Legendre polynomial of order n
Q_n	Legendre function of order n
H_n	Hermite polynomial of order n
L_n	Laguerre polynomial of order n
J_ν	Bessel function of the first kind of order ν
Y_ν	Bessel function of the second kind of order ν
$I_\nu,\ K_\nu$	modified Bessel functions of order ν
Γ	gamma function
H	Heaviside function

Chapter 1
Inner Product Space

An inner product space is the natural generalization of the Euclidean space $\mathbb{R}^n$, with its well-known topological and geometric properties. It constitutes the framework, or setting, for much of our work in this book, as it provides the appropriate mathematical structure that we need. Of particular relevance to this study is the subject of sequences of functions and their integrals, hence a quick review of the convergence properties of such sequences is also included.

1.1 Vector Space

We use the symbol $\mathbb{F}$ to denote either the real number field $\mathbb{R}$ or the complex number field $\mathbb{C}$.

Definition 1.1 A *linear vector space*, or simply a *vector space*, over $\mathbb{F}$ is a set X on which two operations, *addition*

$$+ : X \times X \to X,$$

and *scalar multiplication*

$$\cdot : \mathbb{F} \times X \to X,$$

are defined such that:

1. X is a commutative group under addition, that is,

    (a) $\mathbf{x} + \mathbf{y} = \mathbf{y} + \mathbf{x}$ for all $\mathbf{x}, \mathbf{y} \in X$,
    (b) $\mathbf{x} + (\mathbf{y} + \mathbf{z}) = (\mathbf{x} + \mathbf{y}) + \mathbf{z}$ for all $\mathbf{x}, \mathbf{y}, \mathbf{z} \in X$,

© The Author(s), under exclusive license to Springer-Verlag London Ltd., part of Springer Nature 2026
M. A. Al-Gwaiz, *Sturm-Liouville Theory and its Applications*, Springer Undergraduate Mathematics Series, https://doi.org/10.1007/978-1-4471-7610-7_1

(c) There is a *zero*, or *null*, *element* $\mathbf{0} \in X$ such that $\mathbf{x} + \mathbf{0} = \mathbf{x}$ for all $\mathbf{x} \in X$.

(d) For each $x \in X$ there is an *additive inverse* $-\mathbf{x} \in X$ such that $\mathbf{x} + (-\mathbf{x}) = \mathbf{0}$.

2. Scalar multiplication between the elements of $\mathbb{F}$ and X satisfies

(a) $a \cdot (b \cdot \mathbf{x}) = (ab) \cdot \mathbf{x}$ for all $a, b \in \mathbb{F}$ and all $\mathbf{x} \in X$,

(b) $1 \cdot \mathbf{x} = \mathbf{x}$ for all $\mathbf{x} \in X$.

3. The two distributive properties

(a) $a \cdot (\mathbf{x} + \mathbf{y}) = a \cdot \mathbf{x} + a \cdot \mathbf{y}$,

(b) $(a + b) \cdot \mathbf{x} = a \cdot \mathbf{x} + b \cdot \mathbf{x}$,

  hold for any $a, b \in \mathbb{F}$ and $\mathbf{x}, \mathbf{y} \in X$.

X is called a *real vector space* or a *complex vector space* depending on whether $\mathbb{F} = \mathbb{R}$ or $\mathbb{F} = \mathbb{C}$. The elements of X are called *vectors* and those of $\mathbb{F}$ *scalars*.

From these properties it can be shown that the zero vector $\mathbf{0}$ is unique, and that every $\mathbf{x} \in X$ has a unique additive inverse $-\mathbf{x}$. Furthermore, it follows that $0 \cdot \mathbf{x} = \mathbf{0}$ and $(-1) \cdot \mathbf{x} = -\mathbf{x}$ for every $\mathbf{x} \in X$, and that $a \cdot \mathbf{0} = \mathbf{0}$ for every $a \in \mathbb{F}$. As usual, we often drop the multiplication dot in $a \cdot \mathbf{x}$ and write $a\mathbf{x}$.

Example 1.2

(i) The set of n-tuples of real numbers

$$\mathbb{R}^n = \{(x_1, \cdots, x_n) : x_i \in \mathbb{R}\},$$

under addition, defined by

$$(x_1, \cdots, x_n) + (y_1, \cdots, y_n) = (x_1 + y_1, \cdots, x_n + y_n),$$

and scalar multiplication, defined by

$$a \cdot (x_1, \cdots, x_n) = (ax_1, \cdots, ax_n), \quad a \in \mathbb{R},$$

is a real vector space.

(ii) The set of n-tuples of complex numbers

$$\mathbb{C}^n = \{(z_1, \cdots, z_n) : z_i \in \mathbb{C}\},$$

on the other hand, under the operations

$$(z_1, \cdots, z_n) + (w_1, \cdots, w_n) = (z_1 + w_1, \cdots, z_n + w_n),$$

$$a \cdot (z_1, \cdots, z_n) = (az_1, \cdots, az_n), \quad a \in \mathbb{C},$$

is a complex vector space.

(iii) The set $\mathbb{C}^n$ over the field $\mathbb{R}$ is a real vector space.

(iv) Let I be a real interval which may be closed, open, half-open, finite, or infinite. $\mathcal{P}(I)$ will denote the set of polynomials on I with real (complex) coefficients. This becomes a real (complex) vector space under the usual operation of addition of polynomials, and scalar multiplication

$$b \cdot (a_n x^n + \cdots + a_1 x + a_0) = b a_n x^n + \cdots + b a_1 x + b a_0,$$

where b is a real (complex) number. We also abbreviate $\mathcal{P}(\mathbb{R})$ to $\mathcal{P}$.

(v) The set of real (complex) continuous functions on the real interval I, which will be denoted by $C(I)$, is a real (complex) vector space under the usual operations of addition of functions and multiplication of a function by a real (complex) number.

Let $\{\mathbf{x}_1, \cdots, \mathbf{x}_n\}$ be any finite set of vectors in a vector space X. The sum

$$a_1 \mathbf{x}_1 + \cdots + a_n \mathbf{x}_n = \sum_{i=1}^{n} a_i \mathbf{x}_i, \ \ a_i \in \mathbb{F},$$

is called a *linear combination* of the vectors in the set, and the scalars a_i are the *coefficients* in the linear combination.

Definition 1.3

(i) A finite set of vectors $\{\mathbf{x}_1, \cdots, \mathbf{x}_n\}$ is said to be *linearly independent* if

$$\sum_{i=1}^{n} a_i \mathbf{x}_i = \mathbf{0} \Rightarrow a_i = 0 \ \text{ for all } i \in \{1, \cdots, n\},$$

that is, if every linear combination of the vectors is not equal to zero except when all the coefficients are zeros. The set $\{\mathbf{x}_1, \cdots, \mathbf{x}_n\}$ is *linearly dependent* if it is not linearly independent, i.e., if there is a collection of coefficients $a_1, \cdots, a_n$, not all zeros, such that $\sum_{i=1}^{n} a_i \mathbf{x}_i = \mathbf{0}$.

(ii) An infinite set of vectors $\{\mathbf{x}_1, \mathbf{x}_2, \mathbf{x}_3, \cdots\}$ is *linearly independent* if every finite subset of the set is linearly independent. It is *linearly dependent* if it is not linearly independent, i.e., if there is a finite subset of $\{\mathbf{x}_1, \mathbf{x}_2, \mathbf{x}_3, \cdots\}$ which is linearly dependent.

It should be noted at this point that a finite set of vectors is linearly dependent if, and only if, one of the vectors can be represented as a linear combination of the others (see Exercise 1.3).

Definition 1.4 Let X be a vector space.

(i) A set $\mathcal{A}$ of vectors in X is said to *span* X if every vector in X can be expressed as a linear combination of elements of $\mathcal{A}$. If, in addition, the vectors in $\mathcal{A}$ are linearly independent, then $\mathcal{A}$ is called a *basis* of X.

(ii) A subset Y of X is called a *subspace* of X if every linear combination of vectors in Y lies in Y. This is equivalent to saying that Y is a vector space in its own right (over the same scalar field as X).

If X has a finite basis then any other basis of X is also finite, and both bases have the same number of elements (Exercise 1.4). This number is called the *dimension* of X and is denoted by $\dim X$. If the basis is infinite, we take $\dim X = \infty$.

In Example 1.2, the vectors

$$\mathbf{e}_1 = (1, 0, \cdots, 0),$$

$$\mathbf{e}_2 = (0, 1, 0, \cdots, 0),$$

$$\vdots$$

$$\mathbf{e}_n = (0, \cdots, 0, 1)$$

form a basis for $\mathbb{R}^n$ over $\mathbb{R}$ and $\mathbb{C}^n$ over $\mathbb{C}$. The vectors

$$\mathbf{d}_1 = (i, 0, \cdots, 0),$$

$$\mathbf{d}_2 = (0, i, 0, \cdots, 0),$$

$$\vdots$$

$$\mathbf{d}_n = (0, \cdots, 0, i),$$

together with $\mathbf{e}_1, \cdots, \mathbf{e}_n$, form a basis of $\mathbb{C}^n$ over $\mathbb{R}$. On the other hand, the powers of $x \in \mathbb{R}$,

$$1,\ x,\ x^2,\ x^3, \cdots,$$

span $\mathcal{P}$ and, being linearly independent (Exercise 1.5), they form a basis for the space of real (complex) polynomials over $\mathbb{R}$ ($\mathbb{C}$). Thus both real $\mathbb{R}^n$ and complex $\mathbb{C}^n$ have dimension n, whereas real $\mathbb{C}^n$ has dimension $2n$. The space of polynomials, on the other hand, has infinite dimension. So does the space of continuous functions $C(I)$, as it includes all the polynomials on I (Exercise 1.6).

Let $\mathcal{P}_n(I)$ be the vector space of polynomials on the interval I of degree $\leq n$. This is clearly a subspace of $\mathcal{P}(I)$ of dimension $n + 1$. Similarly, if we denote the set of (real or complex) functions on I whose first derivatives are continuous by $C^1(I)$, then, under the usual operations of addition of functions and multiplication by scalars, $C^1(I)$ is a vector subspace of $C(I)$ over the same (real or complex) field.

As usual, when I is closed at one of its end-points, the derivative at that end-point is the one-sided derivative. More generally, by defining

$$C^n(I) = \{f \in C(I) : f^{(n)} \in C(I)\}, n \in \mathbb{N},$$

$$C^\infty(I) = \bigcap_{n=1}^\infty C^n(I),$$

we obtain a sequence of vector spaces

$$C(I) \supset C^1(I) \supset C^2(I) \supset \cdots \supset C^\infty(I)$$

such that $C^k(I)$ is a (proper) vector subspace of $C^m(I)$ whenever $k > m$. Here $\mathbb{N}$ is the set of natural numbers $\{1, 2, 3, \cdots\}$ and $\mathbb{N}_0 = \mathbb{N} \cup \{0\}$. The set of integers $\{\cdots, -2, -1, 0, 1, 2, \cdots\}$ will be denoted by $\mathbb{Z}$. If we identify $C^0(I)$ with $C(I)$, all the spaces $C^n(I)$, $n \in \mathbb{N}_0$, have infinite dimensions as each includes the polynomials $\mathcal{P}(I)$. When $I = \mathbb{R}$, or when I is not relevant, we simply write C^n.

Exercises

1.1 Use the properties of the vector space X over $\mathbb{F}$ to prove the following:

    (a) $0 \cdot \mathbf{x} = \mathbf{0}$ for all $\mathbf{x} \in X$.
    (b) $a \cdot \mathbf{0} = \mathbf{0}$ for all $a \in \mathbb{F}$.
    (c) $(-1) \cdot \mathbf{x} = -\mathbf{x}$ for all $\mathbf{x} \in X$.
    (d)   If $a \cdot \mathbf{x} = \mathbf{0}$ then either $a = 0$ or $\mathbf{x} = \mathbf{0}$.

1.2 Determine which of the following sets is a vector space under the usual operations of addition and scalar multiplication, and whether it is a real or a complex vector space:

    (a) $\mathcal{P}_n(I)$ with complex coefficients over $\mathbb{C}$.
    (b) $\mathcal{P}(I)$ with imaginary coefficients over $\mathbb{R}$.
    (c)   The set of real numbers over $\mathbb{C}$.
    (d)   The set of complex functions of class $C^n(I)$ over $\mathbb{R}$.

1.3 Prove that the vectors $\mathbf{x}_1, \cdots, \mathbf{x}_n$ are linearly dependent if, and only if, there is an integer $k \in \{1, \cdots, n\}$ such that

$$\mathbf{x}_k = \sum_{i \neq k} a_i \mathbf{x}_i, \ a_i \in \mathbb{F}.$$

Conclude from this that any set of vectors, whether finite or infinite, is linearly dependent if, and only if, one of its vectors is a finite linear combination of the other vectors.

1.4 Let X be a vector space. Prove that, if $\mathcal{A}$ and $\mathcal{B}$ are bases of X and one of them
     is finite, then so is the other and they have the same number of elements.
1.5 Show that the finite set of powers of x, $\{1, x, x^2, \cdots, x^n : x \in (a, b)\}$, is
     linearly independent. Hence conclude that the infinite set $\{1, x, x^2, \cdots : x \in I\}$
     is linearly independent.
1.6 If Y is a subspace of the vector space X, prove that $\dim Y \leq \dim X$.
1.7 Prove that the vectors

$$\mathbf{x}_1 = (x_{11}, \cdots, x_{1n}),$$

$$\vdots$$

$$\mathbf{x}_n = (x_{n1}, \cdots, x_{nn}),$$

where $x_{ij} \in \mathbb{R}$, are linearly dependent if, and only if, $\det(x_{ij}) = 0$, where
$\det(x_{ij})$ is the determinant of the matrix (x_{ij}).

1.2 Inner Product Space

Definition 1.5 Let X be a vector space over $\mathbb{F}$. A function from $X \times X$ to $\mathbb{F}$ is
called an *inner product* in X if, for any pair of vectors $\mathbf{x}, \mathbf{y} \in X$, the inner product
$(\mathbf{x}, \mathbf{y}) \mapsto \langle \mathbf{x}, \mathbf{y} \rangle \in \mathbb{F}$ satisfies the following conditions:

(i) $\langle \mathbf{x}, \mathbf{y} \rangle = \overline{\langle \mathbf{y}, \mathbf{x} \rangle}$ for all $\mathbf{x}, \mathbf{y} \in X$,

(ii) $\langle a\mathbf{x} + b\mathbf{y}, \mathbf{z} \rangle = a \langle \mathbf{x}, \mathbf{z} \rangle + b \langle \mathbf{y}, \mathbf{z} \rangle$ for all $a, b \in \mathbb{F}$, $\mathbf{x}, \mathbf{y}, \mathbf{z} \in X$,

(iii) $\langle \mathbf{x}, \mathbf{x} \rangle \geq 0$ for all $\mathbf{x} \in X$,

(iv) $\langle \mathbf{x}, \mathbf{x} \rangle = 0 \Leftrightarrow \mathbf{x} = \mathbf{0}$.

A vector space on which an inner product is defined is called an *inner product space*.

The symbol $\overline{\langle \mathbf{y}, \mathbf{x} \rangle}$ in (i) denotes the complex conjugate of $\langle \mathbf{y}, \mathbf{x} \rangle$, so that $\langle \mathbf{x}, \mathbf{y} \rangle = \langle \mathbf{y}, \mathbf{x} \rangle$ if X is a real vector space. Note also that (i) and (ii) imply

$$\langle \mathbf{x}, a\mathbf{y} \rangle = \overline{\langle a\mathbf{y}, \mathbf{x} \rangle} = \bar{a} \langle \mathbf{x}, \mathbf{y} \rangle,$$

which means that the linearity property which holds in the first component of the
inner product, as expressed by (ii), does not apply to the second component unless
$\mathbb{F} = \mathbb{R}$.

Theorem 1.6 (Cauchy-Bunyakowsky-Schwarz Inequality) *If X is an inner
product space, then*

$$|\langle \mathbf{x}, \mathbf{y} \rangle|^2 \leq \langle \mathbf{x}, \mathbf{x} \rangle \langle \mathbf{y}, \mathbf{y} \rangle \;\; \textit{for all } \mathbf{x}, \mathbf{y} \in X.$$

Proof If either $\mathbf{x} = \mathbf{0}$ or $\mathbf{y} = \mathbf{0}$ this inequality clearly holds, so we need only consider the case where $\mathbf{x} \neq \mathbf{0}$ and $\mathbf{y} \neq \mathbf{0}$. Furthermore, neither side of the inequality is affected if we replace $\mathbf{x}$ by $a\mathbf{x}$ where $|a| = 1$. Choose a so that $\langle a\mathbf{x}, \mathbf{y} \rangle$ is a real number, that is, if $\langle \mathbf{x}, \mathbf{y} \rangle = |\langle \mathbf{x}, \mathbf{y} \rangle| e^{i\theta}$, let $a = e^{-i\theta}$. Therefore we may assume, without loss of generality, that $\langle \mathbf{x}, \mathbf{y} \rangle$ is a real number. Using the above properties of the inner product, we have, for any real number t,

$$0 \leq \langle \mathbf{x} + t\mathbf{y}, \mathbf{x} + t\mathbf{y} \rangle = \langle \mathbf{x}, \mathbf{x} \rangle + 2 \langle \mathbf{x}, \mathbf{y} \rangle\, t + \langle \mathbf{y}, \mathbf{y} \rangle\, t^2. \tag{1.1}$$

This is a real quadratic expression in t which achieves its minimum at $t = -\langle \mathbf{x}, \mathbf{y} \rangle / \langle \mathbf{y}, \mathbf{y} \rangle$. Substituting this value for t into 1.1 gives

$$0 \leq \langle \mathbf{x}, \mathbf{x} \rangle - \frac{\langle \mathbf{x}, \mathbf{y} \rangle^2}{\langle \mathbf{y}, \mathbf{y} \rangle},$$

and hence the desired inequality. $\qquad\qquad\square$

We now define the *norm* of the vector $\mathbf{x}$ in the inner product space X as

$$\|\mathbf{x}\| = \sqrt{\langle \mathbf{x}, \mathbf{x} \rangle}.$$

Hence, in view of (iii) and (iv), $\|\mathbf{x}\| \geq 0$ for all $\mathbf{x} \in X$, and $\|\mathbf{x}\| = 0$ if and only if $\mathbf{x} = \mathbf{0}$. The Cauchy-Bunyakowsky-Schwarz inequality, which we shall henceforth refer to as the CBS inequality, then takes the form

$$|\langle \mathbf{x}, \mathbf{y} \rangle| \leq \|\mathbf{x}\|\, \|\mathbf{y}\| \quad \text{for all } \mathbf{x}, \mathbf{y} \in X. \tag{1.2}$$

Corollary 1.7 *If X is an inner product space, then*

$$\|\mathbf{x} + \mathbf{y}\| \leq \|\mathbf{x}\| + \|\mathbf{y}\| \quad \textit{for all } \mathbf{x}, \mathbf{y} \in X. \tag{1.3}$$

Proof By definition of the norm,

$$\begin{aligned}
\|\mathbf{x} + \mathbf{y}\|^2 &= \langle \mathbf{x} + \mathbf{y}, \mathbf{x} + \mathbf{y} \rangle \\
&= \|\mathbf{x}\|^2 + \langle \mathbf{x}, \mathbf{y} \rangle + \langle \mathbf{y}, \mathbf{x} \rangle + \|\mathbf{y}\|^2 \\
&= \|\mathbf{x}\|^2 + 2\operatorname{Re} \langle \mathbf{x}, \mathbf{y} \rangle + \|\mathbf{y}\|^2 .
\end{aligned}$$

But $\operatorname{Re} \langle \mathbf{x}, \mathbf{y} \rangle \leq |\langle \mathbf{x}, \mathbf{y} \rangle| \leq \|\mathbf{x}\|\, \|\mathbf{y}\|$ by the CBS inequality, hence

$$\begin{aligned}
\|\mathbf{x} + \mathbf{y}\|^2 &\leq \|\mathbf{x}\|^2 + 2\|\mathbf{x}\|\, \|\mathbf{y}\| + \|\mathbf{y}\|^2 \\
&= (\|\mathbf{x}\| + \|\mathbf{y}\|)^2.
\end{aligned}$$

The inequality (1.3) now follows by taking the square roots of both sides. $\qquad\square$

In $\mathbb{R}^2$ the non-negative number $\|\mathbf{x} - \mathbf{y}\|$ measures the *distance* between the points $\mathbf{x}$ and $\mathbf{y}$ in the plane, and the inequalty

$$\|\mathbf{x} - \mathbf{y}\| = \|\mathbf{x} - \mathbf{z} + \mathbf{z} - \mathbf{y}\|$$
$$\leq \|\mathbf{x} - \mathbf{z}\| + \|\mathbf{z} - \mathbf{y}\|,$$

is the familiar *triangle inequality* which holds between the sides of a triangle whose vertices are the points $\mathbf{x}, \mathbf{y}, \mathbf{z}$. In a general inner product space X, the inner product serves to define the notion of "distance" in this more abstract setting. In mathematical jargon, the space X is now said to be a *topological space*, in which the topology is defined by the norm $\|\cdot\|$, which is derived from the inner product $\langle \cdot, \cdot \rangle$.

Example 1.8

(a) In $\mathbb{R}^n$ we define the inner product of the vectors

$$\mathbf{x} = (x_1, \cdots, x_n), \ \mathbf{y} = (y_1, \cdots, y_n)$$

by

$$\langle \mathbf{x}, \mathbf{y} \rangle = x_1 y_1 + \cdots + x_n y_n, \tag{1.4}$$

which implies

$$\|\mathbf{x}\| = \sqrt{x_1^2 + \cdots + x_n^2}.$$

In this topology the vector space $\mathbb{R}^n$ is the familiar n-dimensional *Euclidean space*. Note that there are other choices for defining the inner product $\langle \mathbf{x}, \mathbf{y} \rangle$, such as $c_1 x_1 y_1 + \cdots + c_n x_n y_n$ where $c_i > 0$ for every i. The provisions of Definition 1.5 would all be satisfied, but the resulting inner product space would not in general be Euclidean.

(b) In $\mathbb{C}^n$ we define

$$\langle \mathbf{z}, \mathbf{w} \rangle = z_1 \bar{w}_1 + \cdots + z_n \bar{w}_n \tag{1.5}$$

for any pair $\mathbf{z}, \mathbf{w} \in \mathbb{C}^n$. Consequently,

$$\|\mathbf{z}\| = \sqrt{|z_1|^2 + \cdots + |z_n|^2}.$$

(c) A natural choice for the definition of an inner product in $C([a, b])$, by analogy with (1.5), is

$$\langle f, g \rangle = \int_a^b f(x)\overline{g(x)}dx, \ \ f, g \in C([a, b]), \tag{1.6}$$

so that

$$\|f\| = \left[\int_a^b |f(x)|^2 \, dx \right]^{1/2}.$$

It is a simple matter to verify that the properties (i), (ii), (iii), and (iv) of the inner product are satisfied in each case, provided of course that $\mathbb{F} = \mathbb{C}$ when the vector space is $\mathbb{C}^n$ or complex $C([a, b])$. To check (iv) in Example 1.8(c), we have to show that

$$\left[\int_a^b |f(x)|^2 \, dx \right]^{1/2} = 0 \quad \Leftrightarrow \quad f(x) = 0 \text{ for all } x \in [a, b].$$

We need only to verify the forward implication ($\Rightarrow$), as the backward implication ($\Leftarrow$) is trivial. But this follows from a well known property of continuous, non-negative functions: If φ is continuous on $[a, b]$, $\varphi \geq 0$, and $\int_a^b \varphi(x)dx = 0$, then $\varphi = 0$ (see [1], for example). Since $|f|^2$ is continuous and non-negative on $[a, b]$ for any $f \in C([a, b])$,

$$\|f\| = 0 \Rightarrow \int_a^b |f(x)|^2 \, dx = 0 \Rightarrow |f|^2 = 0 \Rightarrow f = 0.$$

In this study, we are mainly concerned with function spaces on which an inner product of the type (1.6) is defined. In addition to the topological structure which derives from the norm $\|\cdot\|$, this inner product endows the space with a geometric structure which extends some desirable notions, such as orthogonality, from Euclidean space to infinite dimensional spaces. This will be taken up in Sect. 1.4. Here we examine the Euclidean space $\mathbb{R}^n$ more closely.

Although we proved the CBS and the triangle inequalities for any inner product in Theorem 1.6 and its corollary, we can also derive these inequalities directly in $\mathbb{R}^n$. Consider the inequality

$$(a - b)^2 = a^2 - 2ab + b^2 \geq 0 \tag{1.7}$$

which holds for any pair of real numbers a and b. Let

$$a = \frac{x_i}{\sqrt{x_1^2 + \cdots + x_n^2}}, \quad b = \frac{y_i}{\sqrt{y_1^2 + \cdots + y_n^2}}, \quad x_i, y_i \in \mathbb{R}.$$

If $\sum_{j=1}^{n} x_j^2 \neq 0$ and $\sum_{j=1}^{n} y_j^2 \neq 0$, then (1.7) implies

$$\frac{x_i y_i}{\sqrt{\sum x_j^2}\sqrt{\sum y_j^2}} \leq \frac{1}{2}\frac{x_i^2}{\sum x_j^2} + \frac{1}{2}\frac{y_i^2}{\sum y_j^2},$$

where the summation over the index j is from 1 to n. After summing on i from 1 to n, the right-hand side of this inequality reduces to 1, and we obtain

$$\sum x_i y_i \leq \sqrt{\sum x_i^2} \sqrt{\sum y_i^2}.$$

This inequality remains valid regardless of the signs of x_i and y_i, so we can write

$$\left| \sum x_i y_i \right| \leq \sqrt{\sum x_i^2} \sqrt{\sum y_i^2}$$

for all $\mathbf{x} = (x_1, \cdots, x_n) \neq \mathbf{0}$ and $\mathbf{y} = (y_1, \cdots, y_n) \neq \mathbf{0}$ in $\mathbb{R}^n$. But because the inequality becomes an equality if either $\|\mathbf{x}\|$ or $\|\mathbf{y}\|$ is 0, this proves the CBS inequality

$$|\langle \mathbf{x}, \mathbf{y} \rangle| \leq \|\mathbf{x}\| \, \|\mathbf{y}\| \quad \text{for all } \mathbf{x}, \mathbf{y} \in \mathbb{R}^n.$$

From this the triangle inequality $\|\mathbf{x} + \mathbf{y}\| \leq \|\mathbf{x}\| + \|\mathbf{y}\|$ immediately follows.

Now we define the angle $\theta \in [0, \pi]$ between any pair of non-zero vectors $\mathbf{x}$ and $\mathbf{y}$ in $\mathbb{R}^n$ by the equation

$$\langle \mathbf{x}, \mathbf{y} \rangle = \|\mathbf{x}\| \, \|\mathbf{y}\| \cos \theta.$$

Since the function $\cos : [0, \pi] \to [-1, 1]$ is injective, this defines the angle θ uniquely and agrees with the usual definition of the angle between $\mathbf{x}$ and $\mathbf{y}$ in both $\mathbb{R}^2$ and $\mathbb{R}^3$. With $\mathbf{x} \neq \mathbf{0}$ and $\mathbf{y} \neq \mathbf{0}$,

$$\langle \mathbf{x}, \mathbf{y} \rangle = 0 \quad \Leftrightarrow \quad \cos \theta = 0,$$

which is the condition for the vectors $\mathbf{x}, \mathbf{y} \in \mathbb{R}^n$ to be orthogonal. Consequently, we adopt the following definition of orthogonality in an inner product space.

Definition 1.9

(i) A pair of non-zero vectors $\mathbf{x}$ and $\mathbf{y}$ in the inner product space X is said to be *orthogonal* if $\langle \mathbf{x}, \mathbf{y} \rangle = 0$, symbolically expressed by writing $\mathbf{x} \perp \mathbf{y}$. A set of non-zero vectors $\mathcal{V}$ in X is *orthogonal* if every pair in $\mathcal{V}$ is orthogonal.

(ii) An orthogonal set $\mathcal{V} \subseteq X$ is said to be *orthonormal* if $\|\mathbf{x}\| = 1$ for every $\mathbf{x} \in \mathcal{V}$.

A typical example of an orthonormal set in the Euclidean space $\mathbb{R}^n$ is given by

$$\mathbf{e}_1 = (1, 0, \cdots, 0),$$

$$\mathbf{e}_2 = (0, 1, \cdots, 0),$$

$$\vdots$$

$$\mathbf{e}_n = (0, \cdots, 0, 1),$$

which, as we have already seen, forms a basis of $\mathbb{R}^n$.

In the special case where $X = C([a, b])$ the vectors are represented by functions. Here any two non-zero functions $f, g \in C([a, b])$ are orthogonal if $\langle f, g \rangle = \int_a^b f(x)\overline{g(x)}dx = 0$, and it then follows that $f/\|f\|$ and $g/\|g\|$ are orthonormal. The three polynomial functions $f_1(x) = 1$, $f_2(x) = x$, $f_3(x) = x^2$ are linearly independent over $[-1, 1]$, but only two pairs are orthogonal with respect to the inner product $\langle f, g \rangle = \int_{-1}^1 f(x)g(x)\,dx$, namely $\{1, x\}$ and $\{x, x^2\}$.

In general, if the (non-zero) vectors

$$\mathbf{x}_1, \mathbf{x}_2, \cdots, \mathbf{x}_n \tag{1.8}$$

in the inner product space X are orthogonal, then they are necessarily linearly independent. To see that, let

$$a_1\mathbf{x}_1 + \cdots + a_n\mathbf{x}_n = \mathbf{0}, \ a_i \in \mathbb{F},$$

and take the inner product of each side of this equation with $\mathbf{x}_k$, $1 \le k \le n$. Since $\langle \mathbf{x}_i, \mathbf{x}_k \rangle = 0$ whenever $i \ne k$, we obtain

$$a_k \langle \mathbf{x}_k, \mathbf{x}_k \rangle = a_k \|\mathbf{x}_k\|^2 = 0, \ k \in \{1, \cdots, n\}$$

$$\Rightarrow a_k = 0 \ \text{for all } k.$$

By dividing each vector in (1.8) by its norm, we obtain the orthonormal set $\{\mathbf{x}_i / \|\mathbf{x}_i\| : 1 \le i \le n\}$.

Going back to Euclidean space, let $\mathbf{x}$ be any vector in $\mathbb{R}^n$. We can therefore represent it in the basis $\{\mathbf{e}_1, \cdots, \mathbf{e}_n\}$ by

$$\mathbf{x} = \sum_{i=1}^n a_i\mathbf{e}_i. \tag{1.9}$$

Taking the inner product of Eq. (1.9) with $\mathbf{e}_k$, and using the orthonormal property of $\{\mathbf{e}_i\}$,

$$\langle \mathbf{x}, \mathbf{e}_k \rangle = a_k, \ \ k \in \{1, \cdots, n\}.$$

This determines the coefficients a_i in (1.9), and means that any vector $\mathbf{x}$ in $\mathbb{R}^n$ is represented by the formula

$$\mathbf{x} = \sum_{i=1}^n \langle \mathbf{x}, \mathbf{e}_i \rangle \mathbf{e}_i.$$

The number $\langle \mathbf{x}, \mathbf{e}_i \rangle$ is called the *projection* of $\mathbf{x}$ on $\mathbf{e}_i$, and $\langle \mathbf{x}, \mathbf{e}_i \rangle \mathbf{e}_i$ is the *projection vector* in the direction of $\mathbf{e}_i$. More generally, if $\mathbf{x}$ and $\mathbf{y} \neq \mathbf{0}$ are any vectors in the inner product space X, then $\langle \mathbf{x}, \mathbf{y}/ \|\mathbf{y}\| \rangle$ is the *projection* of $\mathbf{x}$ on $\mathbf{y}$, and the vector

$$\left\langle \mathbf{x}, \frac{\mathbf{y}}{\|\mathbf{y}\|} \right\rangle \frac{\mathbf{y}}{\|\mathbf{y}\|} = \frac{\langle \mathbf{x}, \mathbf{y} \rangle}{\|\mathbf{y}\|^2} \mathbf{y}$$

is its *projection vector* along $\mathbf{y}$, or in the direction of $\mathbf{y}$.

Conversely, suppose that we have a linearly independent set of vectors $\{\mathbf{x}_1, \cdots, \mathbf{x}_n\}$ in the inner product space X. This tacitly implies that none of them is $\mathbf{0}$. Can we form an orthogonal set out of this set? In what follows we present the so-called *Gram-Schmidt method* for constructing an orthogonal set $\{\mathbf{y}_1, \cdots, \mathbf{y}_n\}$ out of $\{\mathbf{x}_i\}$ having the same number of vectors:
We first choose

$$\mathbf{y}_1 = \mathbf{x}_1.$$

The second vector is obtained from $\mathbf{x}_2$ after subtracting the projection vector of $\mathbf{x}_2$ in the direction of $\mathbf{y}_1$,

$$\mathbf{y}_2 = \mathbf{x}_2 - \frac{\langle \mathbf{x}_2, \mathbf{y}_1 \rangle}{\|\mathbf{y}_1\|^2} \mathbf{y}_1.$$

The third vector is $\mathbf{x}_3$ minus the projections of $\mathbf{x}_3$ in the directions of $\mathbf{y}_1$ and $\mathbf{y}_2$,

$$\mathbf{y}_3 = \mathbf{x}_3 - \frac{\langle \mathbf{x}_3, \mathbf{y}_1 \rangle}{\|\mathbf{y}_1\|^2} \mathbf{y}_1 - \frac{\langle \mathbf{x}_3, \mathbf{y}_2 \rangle}{\|\mathbf{y}_2\|^2} \mathbf{y}_2.$$

We continue in this fashion until the last vector

$$\mathbf{y}_n = \mathbf{x}_n - \frac{\langle \mathbf{x}_n, \mathbf{y}_1 \rangle}{\|\mathbf{y}_1\|^2} \mathbf{y}_1 - \cdots - \frac{\langle \mathbf{x}_n, \mathbf{y}_{n-1} \rangle}{\|\mathbf{y}_{n-1}\|^2} \mathbf{y}_{n-1},$$

and the reader can verify that the set $\{\mathbf{y}_1, \cdots, \mathbf{y}_n\}$ is orthogonal.

Exercises

1.8 Given two vectors $\mathbf{x}$ and $\mathbf{y}$ in the inner product space $\mathbb{C}^2$, under what conditions does the equality $\|\mathbf{x} + \mathbf{y}\|^2 = \|\mathbf{x}\|^2 + \|\mathbf{y}\|^2$ hold? Can this equation hold even if the vectors are not orthogonal?

1.9 Let $\mathbf{x}, \mathbf{y} \in X$, where X is an inner product space.

   (a)   If the vectors $\mathbf{x}$ and $\mathbf{y}$ are linearly independent, prove that $\mathbf{x} + \mathbf{y}$ and $\mathbf{x} - \mathbf{y}$ are also linearly independent.

(b) If $\mathbf{x}$ and $\mathbf{y}$ are orthogonal and non-zero, when are $\mathbf{x}+\mathbf{y}$ and $\mathbf{x}-\mathbf{y}$ orthogonal?

1.10 Let $\varphi_1(x) = 1$, $\varphi_2(x) = x$, $\varphi_3(x) = x^2$, $-1 \leq x \leq 1$. Use (1.6) to calculate

(a) $\langle \varphi_1, \varphi_2 \rangle$
(b) $\langle \varphi_1, \varphi_3 \rangle$
(c) $\| \varphi_1 - \varphi_2 \|^2$
(d) $\| 2\varphi_1 + 3\varphi_2 \|$.

1.11 Determine all orthogonal pairs on $[0, 1]$ among the functions $\varphi_1(x) = 1$, $\varphi_2(x) = x$, $\varphi_3(x) = \sin 2\pi x$, $\varphi_4(x) = \cos 2\pi x$. What is the largest orthogonal subset of $\{\varphi_1, \varphi_2, \varphi_3, \varphi_4\}$?

1.12 Determine the projection of $f(x) = \cos^2 x$ on each of the functions $f_1(x) = 1$, $f_2(x) = \cos x$, $f_3(x) = \cos 2x$, $-\pi \leq x \leq \pi$.

1.13 Verify that the functions $\varphi_1, \varphi_2, \varphi_3$ in Exercise 1.10 are linearly independent, and use the Gram-Schmidt method to construct a corresponding orthogonal set.

1.14 Prove that the set of functions $\{1, x, |x|\}$ are linearly independent on $[-1, 1]$, and construct a corresponding orthonormal set. Is the given set linearly independent on $[-1, 0]$?

1.15 If the functions f and g are linearly dependent and differentiable on the real interval I, show that the determinant

$$W(f, g)(x) = f(x)\, g'(x) - g(x)\, f'(x) = \begin{vmatrix} f(x) & f'(x) \\ g(x) & g'(x) \end{vmatrix},$$

known as the *Wronskian* of f and g, vanishes identically on I. Now, since

$$\left(\frac{g}{f} \right)' = \frac{fg' - gf'}{f^2},$$

we see that, conversely, the ratio g/f is constant (and hence f and g are linearly dependent) provided $W(f, g)(x) = 0$ for all $x \in I$ and $f(x) \neq 0$ for any $x \in I$. Can you think of a linearly independent pair of differentiable functions whose Wronskian vanishes identically on $[-1, 1]$?

1.16 Verify that the following functions are orthogonal on $[-1, 1]$:

$$\varphi_1(x) = 1, \quad \varphi_2(x) = x^2 - \frac{1}{3}, \quad \varphi_3(x) = \begin{cases} x/|x|, & x \neq 0 \\ 0, & x = 0. \end{cases}$$

Determine the corresponding orthonormal set.

1.17 Determine the values of the coefficients a and b which make the function $x^2 + ax + b$ orthogonal to both $x + 1$ and $x - 1$ on $[0, 1]$.

1.18 Show that $\sin nx$ and $\sin mx$ are orthogonal on $[-\pi, \pi]$ for all integer values of n and m such that $n \neq m$.

1.3   Sequences of Functions

Much of the subject of this book deals with sequences and series of functions, and this section presents the background that we need on their convergence properties. We assume that the reader is familiar with the basic theory of numerical sequences and series which is usually covered in advanced calculus.

Suppose that for each $n \in \mathbb{N}$ we have a (real or complex) function $f_n : I \to \mathbb{F}$ defined on a real interval I. We then say that we have a *sequence of functions* $(f_n : n \in \mathbb{N})$ defined on I. Suppose, furthermore, that, for every fixed $x \in I$, the sequence of *numbers* $(f_n(x) : n \in \mathbb{N})$ converges as $n \to \infty$ to some limit in $\mathbb{F}$. Now we define the function $f : I \to \mathbb{F}$, for each $x \in I$, by

$$f(x) = \lim_{n \to \infty} f_n(x). \tag{1.10}$$

That means, given any positive number ε, there is a positive integer N such that

$$n \geq N \implies |f_n(x) - f(x)| < \varepsilon. \tag{1.11}$$

Note that the number N depends on the point x as much as it depends on ε, hence $N = N(\varepsilon, x)$. The function f defined in Eq. (1.10) is called the *pointwise limit* of the sequence (f_n).

Definition 1.10 A sequence of functions $f_n : I \to \mathbb{F}$ is said to *converge pointwise* to the function $f : I \to \mathbb{F}$, expressed symbolically by

$$\lim_{n \to \infty} f_n = f, \ \lim f_n = f, \ \text{or} \ f_n \to f,$$

if, for every $x \in I$, $\lim_{n \to \infty} f_n(x) = f(x)$.

Example 1.11

(i) Let $f_n(x) = \dfrac{1}{n} \sin nx$, $x \in \mathbb{R}$. Since

$$\lim_{n \to \infty} f_n(x) = \lim_{n \to \infty} \frac{1}{n} \sin nx = 0 \ \text{ for every } x \in \mathbb{R},$$

the pointwise limit of this sequence is the function $f(x) = 0$, $x \in \mathbb{R}$.

(ii) For all $x \in [0, 1]$,

$$f_n(x) = x^n \to \begin{cases} 0, & 0 \leq x < 1 \\ 1, & x = 1, \end{cases}$$

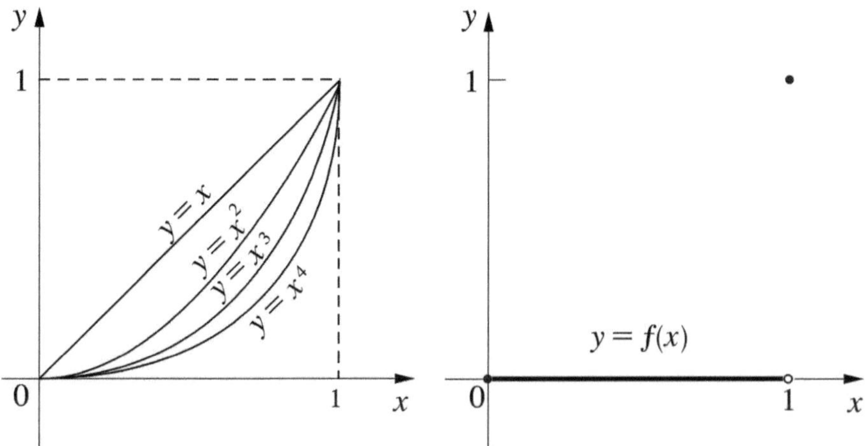

Fig. 1.1 The sequence $f_n(x) = x^n$

hence the limit function is

$$f(x) = \begin{cases} 0, \ 0 \le x < 1 \\ 1, \ x = 1, \end{cases} \tag{1.12}$$

as shown in Fig. 1.1.

(iii) For all $x \in [0, \infty)$, $f_n(x) = \dfrac{nx}{1 + nx} \to f(x) = \begin{cases} 0, \ x = 0 \\ 1, \ x > 0. \end{cases}$

Example 1.12 For each $n \in \mathbb{N}$, define the sequence $f_n : [0, 1] \to \mathbb{R}$ by

$$f_n(x) = \begin{cases} 0, \ x = 0 \\ n, \ 0 < x \le 1/n \\ 0, \ 1/n < x \le 1. \end{cases}$$

To determine the limit f, we first note that $f_n(0) = 0$ for all n. If $x > 0$, then there is an integer N such that $1/N < x$, in which case

$$n \ge N \implies \frac{1}{n} \le \frac{1}{N} < x \implies f_n(x) = 0.$$

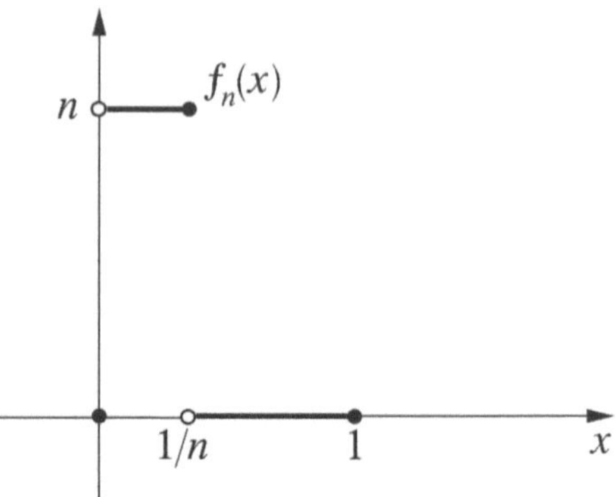

Fig. 1.2 Sequence of functions f_n

Therefore $f_n \to 0$ on $(0, 1]$ (see Fig. 1.2), and the pointwise limit of f_n on $[0, 1]$ is $f = 0$.

If the number N in the implication (1.11) does not depend on x, that is, if for every $\varepsilon > 0$ there is an integer $N = N(\varepsilon)$ such that

$$n \geq N \;\Rightarrow\; |f_n(x) - f(x)| < \varepsilon \;\text{ for all } x \in I, \tag{1.13}$$

then the convergence $f_n \to f$ is called *uniform*, and we distinguish this from pointwise convergence by writing

$$f_n \xrightarrow{u} f.$$

Going back to Example 1.11, we note the following:

(i) Since

$$|f_n(x) - 0| = \left| \frac{1}{n} \sin nx \right| \leq \frac{1}{n} \;\text{ for all } x \in \mathbb{R},$$

we see that any choice of N greater than $1/\varepsilon$ will satisfy the implication (1.13), hence

$$\frac{1}{n} \sin nx \xrightarrow{u} 0 \;\text{ on } \mathbb{R}.$$

(ii) The convergence $x^n \to 0$ is not uniform on $[0, 1)$ because the implication

$$n \geq N \;\Rightarrow\; |x^n - 0| = x^n < \varepsilon$$

cannot be satisfied on the whole interval $[0, 1)$ if $0 < \varepsilon < 1$, but only on $[0, \sqrt[n]{\varepsilon})$, since $x^n > \varepsilon$ for all $x \in (\sqrt[n]{\varepsilon}, 1)$. Hence the convergence $f_n \to f$, where f is given in (1.12), is not uniform.

(iii) The convergence

$$\frac{nx}{1+nx} \to 1, \quad x \in (0, \infty)$$

is also not uniform since the inequality

$$\left|\frac{nx}{1+nx} - 1\right| = \frac{1}{1+nx} < \varepsilon$$

cannot be satisfied for values of x in $(0, (1-\varepsilon)/n\varepsilon]$ if $0 < \varepsilon < 1$.

Remark 1.13

1. The uniform convergence $f_n \overset{u}{\to} f$ clearly implies the pointwise convergence $f_n \to f$, but not vice-versa. Hence, when we wish to test for the uniform convergence of a sequence f_n, the candidate function f for the uniform limit of f_n should always be the pointwise limit.

2. In the inequalities (1.11) and (1.13) we can replace the relation $<$ by $\leq$ and the positive number ε by $c\varepsilon$, where c is a positive constant (which does not depend on n).

3. Since the statement $|f_n(x) - f(x)| \leq \varepsilon$ for all $x \in I$ is equivalent to

$$\sup_{x \in I} |f_n(x) - f(x)| \leq \varepsilon,$$

we see that $f_n \overset{u}{\to} f$ on I if, and only if, for every $\varepsilon > 0$ there is an integer N such that

$$n \geq N \implies \sup_{x \in I} |f_n(x) - f(x)| \leq \varepsilon,$$

which is equivalent to the statement

$$\sup_{x \in I} |f_n(x) - f(x)| \to 0 \ \text{ as } n \to \infty. \tag{1.14}$$

Using the criterion (1.14) for uniform convergence on the sequences of Example 1.11, we see that, in (i),

$$\sup_{x \in \mathbb{R}} \left|\frac{1}{n} \sin nx\right| \leq \frac{1}{n} \to 0,$$

thus confirming the uniform convergence of $\sin nx/n$ to 0. In (ii) and (iii), we have

$$\sup_{x\in[0,1]} \left|x^n - 0\right| = \sup_{x\in[0,1)} x^n = 1 \nrightarrow 0,$$

$$\sup_{x\in[0,\infty)} \left|\frac{nx}{1+nx} - f(x)\right| = \sup_{x\in(0,\infty)} \left(1 - \frac{nx}{1+nx}\right) = 1 \nrightarrow 0,$$

hence neither sequence converges uniformly.

Although all three sequences discussed in Example 1.11 are continuous, only the first one, $(\sin nx/n)$, converges to a continuous limit. This would seem to indicate that uniform convergence preserves the property of continuity as the sequence passes to the limit. We should also be interested to know under what conditions we can interchange the order of integration or differentiation with the process of passage to the limit. In other words, when can we write

$$\int_I \lim f_n(x)dx = \lim \int_I f_n(x)dx, \text{ or } (\lim f_n)' = \lim f_n' \text{ on } I?$$

The answer is given in the following theorem, which gives sufficient conditions for the validity of these equalities. This is a standard result in classical real analysis whose proof may be found in a number of references, such as [1] or [15].

Theorem 1.14 *Let (f_n) be a sequence of functions defined on the interval I which converges pointwise to f on I.*

(i) *If f_n is continuous for every n, and $f_n \overset{u}{\rightarrow} f$, then f is continuous on I.*

(ii) *If f_n is integrable for every n, I is bounded, and $f_n \overset{u}{\rightarrow} f$, then f is integrable on I and*

$$\int_I f(x)dx = \lim \int_I f_n(x)dx.$$

(iii) *If f_n is differentiable on I for every n, I is bounded, and f_n' converges uniformly on I, then f_n converges uniformly to f, f is differentiable on I, and*

$$f_n' \overset{u}{\rightarrow} f' \text{ on } I.$$

Remark 1.15 Part (iii) of Theorem 1.14 remains valid if pointwise convergence of f_n on I is replaced by the weaker condition that f_n converges at any single point in I.

Going back to Example 1.11, we observe that the uniform convergence of $\sin nx/n$ to 0 satisfies part (i) of Theorem 1.14. It also satisfies (ii) over any bounded interval in $\mathbb{R}$. But (iii) is not satisfied, since the sequence

$$\frac{d}{dx}\left(\frac{1}{n}\sin nx\right) = \cos nx$$

is not convergent. The sequence (x^n) is continuous on $[0, 1]$ for every n, but its limit is not. This is consistent with with (i), since the convergence is not uniform. The same observation applies to the sequence $nx/(1 + nx)$.

In Example 1.12 we have

$$\int_0^1 f_n(x)dx = \int_0^{1/n} ndx = 1 \ \text{ for all } n \in \mathbb{N}$$

$$\Rightarrow \lim \int_0^1 f_n(x)dx = 1,$$

whereas

$$\int_0^1 \lim f_n(x)dx = 0.$$

This implies that the convergence $f_n \to 0$ is not uniform, which is confirmed by the fact that

$$\sup_{0 \le x \le 1} f_n(x) = n.$$

On the other hand,

$$\lim \int_0^1 x^n dx = 0 = \int_0^1 \lim x^n dx,$$

though the convergence $x^n \to 0$ is not uniform, which indicates that not all the conditions of Theorem 1.14 are necessary.

Given a sequence of (real or complex) functions (f_n) defined on a real interval I, we define its n-th partial sum by

$$S_n(x) = f_1(x) + \cdots + f_n(x) = \sum_{k=1}^n f_k(x), \ x \in I.$$

The sequence of functions (S_n), defined on I, is called an *infinite series* (of functions) and is denoted by $\sum f_k$. The series is said to converge pointwise on I if the sequence (S_n) converges pointwise on I, in which case $\sum f_k$ is said to be *convergent*. Its limit is the infinite sum of the series

$$\lim_{n \to \infty} S_n(x) = \sum_{k=1}^{\infty} f_k(x), \ x \in I.$$

Sometimes we shall find it convenient to identify a convergent series with its sum, just as we occasionally identify a function f with its value $f(x)$. A series which does not converge at a point is said to *diverge* at that point. The series $\sum f_k$ is *absolutely convergent* on I if the positive series $\sum |f_k|$ is pointwise convergent on I, and *uniformly convergent* on I if the sequence (S_n) is uniformly convergent on I. Uniform convergence just means, given $\varepsilon > 0$, there is a number N (which depends only on ε) such that

$$n \geq N \Rightarrow \left| \sum_{k=1}^{\infty} f_k(x) - \sum_{k=1}^{n} f_k(x) \right| = \left| \sum_{k=n+1}^{\infty} f_k(x) \right| < \varepsilon \ \text{ for all } x \in I.$$

In investigating the convergence properties of series of functions we naturally rely on the corresponding convergence properties of sequences of functions, as discussed earlier, because a series is ultimately a sequence. But we often resort to the convergence properties of series of numbers, which we assume that the reader is familiar with, such as the various tests of convergence (comparison test, ratio test, root test, alternating series test), and the convergence properties of such series as the geometric series $\sum x^n$ and the p-series $\sum 1/n^p$ (see [1] or [2]).

Applying Theorem 1.14 to series, we arrive at the following result.

Corollary 1.16 *Suppose the series $\sum f_n$ converges pointwise on the interval I.*

(i) *If f_n is continuous on I for every n and $\sum f_n$ converges uniformly on I, then its sum $\sum_{n=1}^{\infty} f_n$ is continuous.*

(ii) *If f_n is integrable on I for every n, I is bounded, and $\sum f_n$ converges uniformly, then $\sum_{n=1}^{\infty} f_n$ is integrable on I and*

$$\int_I \sum_{n=1}^{\infty} f_n(x)dx = \sum_{n=1}^{\infty} \int_I f_n(x)dx.$$

(iii) *If f_n is differentiable on I for every n, I is bounded, and $\sum f_n'$ converges uniformly on I, then $\sum f_n$ converges uniformly and its limit is differentiable on I and satisfies*

$$\left(\sum_{n=1}^{\infty} f_n \right)' = \sum_{n=1}^{\infty} f_n'.$$

This corollary points out the relevance of uniform convergence to manipulating series, and it would be helpful if we had a simpler and more practical test for the uniform convergence of a series than applying the definition. This is provided by the following theorem, which gives a sufficient condition for the uniform convergence of a series of functions.

Theorem 1.17 (Weierstrass M-Test) *Let (f_n) be a sequence of functions on I, and suppose that there is a corresponding sequence of numbers M_n such that*

$$|f_n(x)| \leq M_n \ \text{for all } x \in I, n \in \mathbb{N}.$$

If $\sum M_n$ converges, then $\sum f_n$ converges uniformly and absolutely on I.

Proof Let $\varepsilon > 0$. We have

$$\left| \sum_{k=1}^{\infty} f_k(x) - \sum_{k=1}^{n} f_k(x) \right| \leq \sum_{k=n+1}^{\infty} |f_k(x)|$$

$$\leq \sum_{k=n+1}^{\infty} M_k \ \text{ for all } x \in I, n \in \mathbb{N}.$$

Since the series $\sum M_k$ is convergent, there is an integer N such that

$$n \geq N \Rightarrow \sum_{k=n+1}^{\infty} M_k < \varepsilon$$

$$\Rightarrow \left| \sum_{k=1}^{\infty} f_k(x) - \sum_{k=1}^{n} f_k(x) \right| < \varepsilon \ \text{ for all } x \in I.$$

Absolute convergence follows by comparison with $\sum M_n$.   □

Example 1.18

(i) The trigonometric series

$$\sum \frac{1}{n^2} \sin nx$$

is uniformly convergent on $\mathbb{R}$ because

$$\left| \frac{1}{n^2} \sin nx \right| \leq \frac{1}{n^2}$$

and the series $\sum 1/n^2$ is convergent. Since $\sin nx/n^2$ is continuous on $\mathbb{R}$ for every n, the function $\sum_{n=1}^{\infty} \sin nx/n^2$ is also continuous on $\mathbb{R}$. Furthermore, by Corollary 1.16, the integral of the series on any finite interval $[a, b]$ is

$$\int_a^b \left(\sum_{n=1}^{\infty} \frac{1}{n^2} \sin nx \right) dx = \sum_{n=1}^{\infty} \frac{1}{n^2} \int_a^b \sin nx \, dx$$

$$= \sum_{n=1}^{\infty} \frac{1}{n^3} (\cos na - \cos nb).$$

$$\sum_{n=1}^{\infty} \frac{1}{n^3} |\cos na - \cos nb| \leq 2 \sum_{n=1}^{\infty} \frac{1}{n^3},$$

which is convergent, and hence the left-hand side is absolutely and uniformly convergent on $\mathbb{R}$. On the other hand, the series of derivatives

$$\sum_{n=1}^{\infty} \frac{d}{dx} \left(\frac{1}{n^2} \sin nx \right) = \sum_{n=1}^{\infty} \frac{1}{n} \cos nx$$

is not uniformly convergent. In fact, it is divergent at some values of x, such as the integral multiples of 2π. Hence we cannot write

$$\frac{d}{dx} \sum_{n=1}^{\infty} \frac{1}{n^2} \sin nx = \sum_{n=1}^{\infty} \frac{1}{n} \cos nx \quad \text{for all } x \in \mathbb{R}.$$

(ii) By the M-test, both the series

$$\sum \frac{1}{n^3} \sin nx$$

and

$$\sum_{n=1}^{\infty} \frac{d}{dx} \left(\frac{1}{n^3} \sin nx \right) = \sum_{n=1}^{\infty} \frac{1}{n^2} \cos nx$$

are uniformly convergent on $\mathbb{R}$. Hence the equality

$$\frac{d}{dx} \sum_{n=1}^{\infty} \frac{1}{n^3} \sin nx = \sum_{n=1}^{\infty} \frac{1}{n^2} \cos nx$$

is valid for all x in $\mathbb{R}$.

Exercises

1.19 Calculate the pointwise limit as $n \to \infty$ where it exists:

(a) $\dfrac{x^n}{1 + x^n}$, $x \in \mathbb{R}$,

(b) $\sqrt[n]{x}$, $0 \le x < \infty$,

(c) $\sin nx$, $x \in \mathbb{R}$.

1.20 Determine the type of convergence (pointwise or uniform) for each of the following sequences:

(a) $\dfrac{x^n}{1 + x^n}$, $0 \le x \le 2$,

(b) $\sqrt[n]{x}$, $1/2 \le x \le 1$,

(c) $\sqrt[n]{x}$, $0 \le x \le 1$.

1.21 Determine the type of convergence for the sequence

$$f_n(x) = \begin{cases} nx, & 0 \le x < 1/n \\ 1, & 1/n \le x \le 1, \end{cases}$$

and decide whether the equality

$$\lim \int_0^1 f_n(x)dx = \int_0^1 \lim f_n(x)dx$$

is valid.

1.22 Evaluate the limit of the sequence

$$f_n(x) = \begin{cases} nx, & 0 \le x \le 1/n \\ n(1 - x)/(n - 1), & 1/n < x \le 1, \end{cases}$$

and determine its type of convergence. Sketch the function $\lim f_n(x)$.

1.23 Determine the limit and the type of convergence for the sequence $f_n(x) = nx(1 - x^2)^n$ on $[0, 1]$.

1.24 Prove that the convergence

$$\frac{x}{n+x} \to 0$$

is uniform on $[0, a]$ for any $a > 0$, but not on $[0, \infty)$.

1.25 Given

$$f_n(x) = \begin{cases} 1/n, & |x| \le n \\ 0, & |x| > n, \end{cases}$$

prove that $f_n \overset{u}{\to} 0$. Evaluate $\lim \int_{-\infty}^{\infty} f_n(x)dx$ and explain why it is not 0.

1.26 If the sequence (f_n) converges uniformly to f on $[a, b]$, prove that $|f_n - f|$, and hence $|f_n - f|^2$, converges uniformly to 0 on $[a, b]$.

1.27 Determine the interval of convergence of the series $\sum f_n$, where

(a) $f_n(x) = \dfrac{1}{n^2 + x^2}$,

(b) $f_n(x) = \dfrac{x^n}{1 + x^n}$.

1.28 Prove that

$$\lim_{n \to \infty} \int_{n\pi}^{(n+1)\pi} \frac{|\sin x|}{x} dx = 0.$$

Use this to conclude that the improper integral

$$\int_0^{\infty} \frac{\sin x}{x} dx$$

exists. Show that the integral $\int_0^{\infty} (|\sin x|/x)dx = \infty$. Hint: Use the alternating series test and the divergence of the harmonic series $\sum 1/n$.

1.29 Determine the domain of convergence of the series $\sum_{n=0}^{\infty} e^{-nx}$.

1.30 The series

$$\sum_{n=0}^{\infty} a_n x^n = a_0 + a_1 x + a_2 x^2 + \cdots$$

is called a *power series* about the point 0. It is known that this series converges in $(-R, R)$ and diverges outside $[-R, R]$, where

$$R = \left[\limsup \sqrt[n]{|a_n|} \right]^{-1} = \lim_{n \to \infty} \left| \frac{a_n}{a_{n+1}} \right| \ge 0,$$

provided $\lim_{n\to\infty} |a_n/a_{n+1}|$ exists. If $R > 0$ use the Weierstrass M-test to prove that the power series converges uniformly on $[-R + \varepsilon, R - \varepsilon]$, where ε is any positive number less than R.

1.31 Use the result of Exercise 1.30 and to show that the function

$$f(x) = \sum_{n=0}^{\infty} a_n x^n$$

is continuous on $(-R, R)$; then show that f is also differentiable on $(-R, R)$ with

$$f'(x) = \sum_{n=1}^{\infty} n a_n x^{n-1}.$$

1.32 From Exercise 1.31 conclude that the power series $f(x) = \sum_{n=0}^{\infty} a_n x^n$ is differentiable any number of times on $(-R, R)$, and that $a_n = f^{(n)}(0)/n!$ for all $n \in \mathbb{N}$.

1.33 Use the result of Exercise 1.32 and Taylor's theorem to obtain the following power series representations of the exponential and trigonometric functions on $\mathbb{R}$:

$$e^x = \sum_{n=0}^{\infty} \frac{x^n}{n!},$$

$$\cos x = \sum_{n=0}^{\infty} (-1)^n \frac{x^{2n}}{(2n)!},$$

$$\sin x = \sum_{n=0}^{\infty} (-1)^n \frac{x^{2n+1}}{(2n+1)!}.$$

1.34 Use the result of Exercise 1.33 to prove Euler's identity $e^{ix} = \cos x + i \sin x$ for all $x \in \mathbb{R}$, where $i = \sqrt{-1}$.

1.4 The Space $\mathcal{L}^2$

For any two functions f and g in the vector space $C([a, b])$ of complex continuous functions on a real interval $[a, b]$, we defined the inner product

$$\langle f, g \rangle = \int_a^b f(x)\overline{g(x)}dx \tag{1.15}$$

in Sect. 1.2, from which followed the definition of the norm

$$\|f\| = \sqrt{\langle f, f \rangle} = \sqrt{\int_a^b |f(x)|^2 \, dx}. \tag{1.16}$$

As in $\mathbb{R}^n$, we can also show directly that the CBS inequality holds in $C([a, b])$. For any $f, g \in C([a, b])$, we have

$$\left\| \frac{|f|}{\|f\|} - \frac{|g|}{\|g\|} \right\|^2 = \int_a^b \left[\frac{|f(x)|}{\|f\|} - \frac{|g(x)|}{\|g\|} \right]^2 dx \geq 0,$$

where we assume that $\|f\| \neq 0$ and $\|g\| \neq 0$. Hence

$$\int_a^b \frac{|f(x)|}{\|f\|} \frac{|g(x)|}{\|g\|} dx \leq \frac{1}{2\|f\|^2} \int_a^b |f(x)|^2 \, dx + \frac{1}{2\|g\|^2} \int_a^b |g(x)|^2 \, dx = 1$$

$$\Rightarrow \langle |f|, |g| \rangle \leq \|f\| \, \|g\|.$$

Using the monotonicity property of the integral

$$\left| \int_a^b \varphi(x) dx \right| \leq \int_a^b |\varphi(x)| \, dx,$$

we therefore conclude that

$$|\langle f, g \rangle| \leq \langle |f|, |g| \rangle \leq \|f\| \, \|g\|. \tag{1.17}$$

If either $\|f\| = 0$ or $\|g\| = 0$ the inequality remains valid, as it becomes an equality. The triangle inequality

$$\|f + g\| \leq \|f\| + \|g\|$$

then easily follows from the relation $f\bar{g} + \bar{f}g = 2\mathrm{Re} f\bar{g} \leq 2|fg|$.

As we have already observed, the non-negative number $\|f - g\|$ may be regarded as a measure of the "distance" between the functions $f, g \in C([a, b])$. In this case we clearly have $\|f - g\| = 0$ if, and only if, $f = g$ on $[a, b]$. This is the advantage of dealing with continuous functions, for if we admit discontinuous functions, such as

$$h(x) = \begin{cases} 1, & x = 1 \\ 0, & x \in (1, 2], \end{cases} \tag{1.18}$$

then $\|h\| = 0$ whereas $h \neq 0$.

Nevertheless, $C([a, b])$ is not a suitable inner product space for pursuing this study, for it is not closed under pointwise limit operations as we have seen in Sect. 1.3. So we need to enlarge the inner product space of continuous functions over $[a, b]$ in order to overcome this difficulty. Moreover, in this larger space we can only admit functions for which the inner product

$$\langle f, g \rangle = \int_a^b f(x)\overline{g(x)}dx$$

is defined for every pair f, g in the space. But, in spite of the CBS inequality (1.17), the integrability of f^2 and g^2 does not imply the integrabily of fg in the sense of Riemann (see Exercise 1.36). So we have to enlarge the class $\mathcal{R}(a, b)$ of Riemann integrable functions over $[a, b]$ to the corresponding larger class of Lebesgue integrable functions $\mathcal{L}^1(a, b)$, as defined in Section A.2 of the Appendix, where the inequality (1.17) implies that $fg \in \mathcal{L}^1(a, b)$ whenever f^2 and g^2 are integrable. Another advantage to this step is the closure properties that $\mathcal{L}^1(a, b)$ enjoys vis-a-vis $\mathcal{R}(a, b)$ under certain limiting operations, as discussed in Sects. 1.3 and A.2.

We therefore arrive at the space $\mathcal{L}^2(a, b)$ of functions $f : (a, b) \to \mathbb{C}$ such that $f^2 \in \mathcal{L}^1(a, b)$, i.e. the measurable functions f on $[a, b]$ such that

$$\int_a^b |f(x)|^2\, dx < \infty,$$

as the desirable setting for much of our work. For a complex function $f(x) = u(x) + iv(x)$, f is measurable on $[a, b]$ if u and v are measurable, and Lebesgue integrable if both are. Naturally we will be dealing with more general domains of definition for the function f, such as open, semi infinite, and infinite intervals. In theory any measurable set in $\mathbb{R}$ is admissible, but in practice the integrals we encounter will almost exclusively be of the Riemann type over real intervals, both proper and improper. The importance of the more general Lebesgue theory is really in providing the tools to justify certain limiting procedures which cannot be justified in the Riemann theory, on the grounds that any function which is Riemann integrable in the proper sense is also Lebesgue integrable (see Sect. A.2 of Appendix A). By defining the inner product (1.15) and the norm (1.16) on $\mathcal{L}^2(a, b)$, we can use the triangle inequality to obtain

$$\|\alpha f + \beta g\| \leq \|\alpha f\| + \|\beta g\|$$

$$= |\alpha|\, \|f\| + |\beta|\, \|g\| \text{ for all } f, g \in \mathcal{L}^2(a, b), \ \alpha, \beta \in \mathbb{C},$$

hence $\alpha f + \beta g \in \mathcal{L}^2(a, b)$ whenever $f, g \in \mathcal{L}^2(a, b)$. Thus $\mathcal{L}^2(a, b)$ is a linear vector space which, under the inner product (1.15), becomes an inner product space and includes $C([a, b])$ as a proper subspace.

In $\mathcal{L}^2(a, b)$ the equality $\|f\| = 0$ does not necessarily mean $f(x) = 0$ at every point $x \in [a, b]$. For example, in the case where $f(x) = 0$ on all but a finite number of points in $[a, b]$ we clearly have $\|f\| = 0$. The function $f = 0$ pointwise on a real interval I if $f(x) = 0$ at every $x \in I$. If $\|f\| = 0$ we say that $f = 0$ in $\mathcal{L}^2(I)$. Thus the function h defined in (1.18) equals 0 in $\mathcal{L}^2(I)$, but not pointwise. The function 0 in $\mathcal{L}^2(I)$ really denotes an *equivalence class of functions*, each of which has norm 0. The function which is pointwise equal to 0 is only one member, indeed the only continuous member, of that class. Similarly, we say that two functions f and g in $\mathcal{L}^2(I)$ are equal *in* $\mathcal{L}^2(I)$ if $\|f - g\| = 0$, though f and g may not be equal pointwise. In the terminology of measure theory, f and g are said to be "equal almost everywhere", written as $f = g$ a.e. Hence the space $\mathcal{L}^2(I)$ is, in fact, made up of equivalence classes of functions, and each equivalence class is defined by equality almost everywhere. Needless to say, the integral $\int_I f$ of an integrable function over a bounded interval is the same whether I is closed, open, or half-open. This is actually true whether the integral is defined in the sense of Lebesgue or Riemann. When I or the function f is unbounded we interpret the integral on I as an improper integral that can be evaluated by integrating a suitably truncated function f_n, which approaches f, and taking the limit of $\int f_n$ as $n \to \infty$.

Example 1.19 Determine the functions that belong to $\mathcal{L}^2$ and calculate their norms:

(i) $f(x) = \begin{cases} 1, & 0 \le x < 1/2 \\ 0, & 1/2 \le x \le 1. \end{cases}$

(ii) $f(x) = 1/\sqrt{x}, \ 0 < x < 1.$

(iii) $f(x) = 1/\sqrt[3]{x}, \ 0 < x < 1.$

(iv) $f(x) = 1/x, \ 1 < x < \infty.$

Solution

(i)

$$\|f\|^2 = \int_0^1 f^2(x)dx = \int_0^{1/2} dx = \frac{1}{2}.$$

Therefore $f \in \mathcal{L}^2(0, 1)$ and $\|f\| = 1/\sqrt{2}$.

(ii)

$$\|f\|^2 = \int_0^1 \frac{1}{x}dx = \lim_{a \to 0^+} \int_a^1 \frac{1}{x}dx = -\lim_{\varepsilon \to 0^+} \log \varepsilon = \infty$$

$$\Rightarrow f \notin \mathcal{L}^2(0, 1).$$

(iii)

$$\|f\|^2 = \int_0^1 \frac{1}{x^{2/3}}dx = \lim_{\varepsilon \to 0^+} 3(1 - \varepsilon^{1/3}) = 3$$

$$\Rightarrow f \in \mathcal{L}^2(0, 1), \quad \|f\| = \sqrt{3}.$$

(iv)

$$\|f\|^2 = \int_1^\infty \frac{1}{x^2}dx = \lim_{b \to \infty} \int_1^b \frac{1}{x^2}dx = \lim_{b \to \infty}\left(1 - \frac{1}{b}\right) = 1$$

$$\Rightarrow f \in \mathcal{L}^2(1, \infty), \quad \|f\| = 1.$$

Example 1.20 The infinite set of functions $\{1, \cos x, \sin x, \cos 2x, \sin 2x, \cdots\}$ is orthogonal in the real inner product space $\mathcal{L}^2(-\pi, \pi)$. This can be seen by calculating the inner product of each pair in the set:

$$\langle 1, \cos nx \rangle = \int_{-\pi}^{\pi} \cos nx \, dx = 0, \quad n \in \mathbb{N}.$$

$$\langle 1, \sin nx \rangle = \int_{-\pi}^{\pi} \sin nx \, dx = 0, \quad n \in \mathbb{N}.$$

$$\langle \cos nx, \cos mx \rangle = \int_{-\pi}^{\pi} \cos nx \cos mx \, dx$$

$$= \frac{1}{2} \int_{-\pi}^{\pi} [\cos(n - m)x + \cos(n + m)x]dx$$

$$= \frac{1}{2}\left[\frac{1}{n - m}\sin(n - m)x + \frac{1}{n + m}\sin(n + m)x\right]\Big|_{-\pi}^{\pi}$$

$$= 0, \quad n \neq m.$$

$$\langle \sin nx, \sin mx \rangle = \int_{-\pi}^{\pi} \sin nx \sin mx \, dx$$

$$= \frac{1}{2} \int_{-\pi}^{\pi} [\cos(n - m)x - \cos(n + m)x]dx$$

$$= 0, \quad n \neq m.$$

$$\langle \cos nx, \sin mx \rangle = \int_{-\pi}^{\pi} \cos nx \sin mx \, dx = 0, \quad n, m \in \mathbb{N}.$$

Furthermore,

$$\|1\| = \sqrt{2\pi},$$

$$\|\cos nx\| = \left[\int_{-\pi}^{\pi} \cos^2 nx \, dx\right]^{1/2} = \sqrt{\pi},$$

$$\|\sin nx\| = \left[\int_{-\pi}^{\pi} \sin^2 nx \, dx\right]^{1/2} = \sqrt{\pi}, \quad n \in \mathbb{N}.$$

Thus the set

$$\left\{\frac{1}{\sqrt{2\pi}}, \frac{\cos x}{\sqrt{\pi}}, \frac{\sin x}{\sqrt{\pi}}, \frac{\cos 2x}{\sqrt{\pi}}, \frac{\sin 2x}{\sqrt{\pi}}, \cdots\right\},$$

which is obtained by dividing each function in the orthogonal set by its norm, is orthonormal in $\mathcal{L}^2(-\pi, \pi)$.

Example 1.21 The set of functions

$$\{e^{inx} : n \in \mathbb{Z}\} = \{\cdots, e^{-i2x}, e^{-ix}, 1, e^{ix}, e^{i2x}, \cdots\}$$

is orthogonal in the complex space $\mathcal{L}^2(-\pi, \pi)$, since, for any $n \neq m$,

$$\left\langle e^{inx}, e^{imx} \right\rangle = \int_{-\pi}^{\pi} e^{inx} \overline{e^{imx}} dx$$

$$= \int_{-\pi}^{\pi} e^{inx} e^{-imx} dx$$

$$= \frac{1}{i(n-m)} e^{i(n-m)x} \Big|_{-\pi}^{\pi} = 0.$$

By dividing the functions in this set by

$$\left\|e^{inx}\right\| = \left[\int_{-\pi}^{\pi} e^{inx} \overline{e^{inx}} dx\right]^{1/2} = \sqrt{2\pi}, \quad n \in \mathbb{Z},$$

we obtain the corresponding orthonormal set $\left\{\dfrac{1}{\sqrt{2\pi}} e^{inx} : n \in \mathbb{Z}\right\}$.

If ρ is a positive continuous function on (a, b), we can also define the inner product of two functions f and g with respect to the *weight function* ρ by

$$\langle f, g \rangle_\rho = \int_a^b f(x)\overline{g(x)}\rho(x)dx, \tag{1.19}$$

and we leave it to the reader to verify that all the properties of the inner product, as given in Definition 1.5, are satisfied. f is then said to be orthogonal to g with respect to the weight function ρ if $\langle f, g \rangle_\rho = 0$. The induced norm

$$\|f\|_\rho = \left[\int_a^b |f(x)|^2 \, \rho(x) dx \right]^{1/2}$$

satisfies all the properties of the norm (1.16), including the CBS inequality and the triangle inequality. We shall use $\mathcal{L}^2_\rho(a, b)$ to denote the set of functions $f : (a, b) \to \mathbb{C}$, where (a, b) may be finite or infinite, such that $\|f\|_\rho < \infty$. This is clearly an inner product space, and $\mathcal{L}^2(a, b)$ is then the special case in which $\rho \equiv 1$.

Exercises

1.35 Verify the CBS inequality for the functions $f(x) = 1$ and $g(x) = x$ on $[0, 1]$.

1.36 Let the functions f and g be defined on $[0, 1]$ by

$$f(x) = \begin{cases} 1, & x \in \mathbb{Q} \cap [0, 1] \\ -1, & x \in \mathbb{Q}^c \cap [0, 1] \end{cases}, \quad g(x) = 1 \text{ for all } x \in [0, 1],$$

where $\mathbb{Q}$ is the set of rational numbers and $\mathbb{Q}^c$ is its complement in $\mathbb{R}$, i.e the irrational real numbers, also denoted by $\mathbb{R} \backslash \mathbb{Q}$. Show that both f^2 and g^2 are Riemann integrable on $[0, 1]$ but that fg is not.

1.37 Determine which of the following functions belongs to $\mathcal{L}^2(0, \infty)$ and calculate its norm:

$$(i)\ e^{-x}, \ (ii)\ \sin x, \ (iii)\ \frac{1}{1+x}, \ (iv)\ \frac{1}{\sqrt[3]{x}}.$$

1.38 If f and g are positive, continuous functions in $\mathcal{L}^2(a, b)$, prove that $\langle f, g \rangle = \|f\| \, \|g\|$ if, and only if, f and g are linearly dependent.

1.39 Discuss the conditions under which the equality $\|f + g\| = \|f\| + \|g\|$ holds in $\mathcal{L}^2(a, b)$.

1.40 Determine the real values of α for which x^α lies in $\mathcal{L}^2(0, 1)$.

1.41 Determine the real values of α for which x^α lies in $\mathcal{L}^2(1, \infty)$.

1.42 If $f \in \mathcal{L}^2(0, \infty)$ and $\lim_{x \to \infty} f(x)$ exists, prove that $\lim_{x \to \infty} f(x) = 0$.

1.43 Assuming the interval (a, b) is finite, prove that $f \in \mathcal{L}^2(a, b) \subset \mathcal{L}^1(a, b)$. Show that the converse is false by giving an example of a function f such that $|f|$ is integrable on (a, b), but $f \notin \mathcal{L}^2(a, b)$.

1.44 If the function $f : [0, \infty) \to \mathbb{R}$ is bounded and $|f|$ is integrable, prove that $f \in \mathcal{L}^2(0, \infty)$. Show that the converse is false by giving an example of a bounded function in $\mathcal{L}^2(0, \infty)$ which is not integrable on $[0, \infty)$.

1.45 In $\mathcal{L}^2(-\pi, \pi)$, express the function $\sin^3 x$ as a linear combination of the orthogonal functions $\{1, \cos x, \sin x, \cos 2x, \sin 2x, \cdots\}$.

1.46 Given $\rho(x) = e^{-x}$, prove that any polynomial in x belongs to $\mathcal{L}_\rho^2(0, \infty)$.

1.47 Show that if ρ and σ are two weight functions such that $\rho \geq \sigma \geq 0$ on (a, b), then $\mathcal{L}_\rho^2(a, b) \subseteq \mathcal{L}_\sigma^2(a, b)$.

1.5   Convergence in $\mathcal{L}^2$

Having discussed pointwise and uniform convergence for a sequence of functions, we now consider a third type: convergence in $\mathcal{L}^2$.

Definition 1.22 A sequence of functions (f_n) in $\mathcal{L}^2(a, b)$ is said to *converge in $\mathcal{L}^2$* if there is a function $f \in \mathcal{L}^2(a, b)$ such that

$$\lim_{n \to \infty} \| f_n - f \| = 0, \tag{1.20}$$

that is, if for every $\varepsilon > 0$ there is an integer N such that

$$n \geq N \Rightarrow \| f_n - f \| < \varepsilon.$$

Equation (1.20) is equivalent to writing

$$f_n \xrightarrow{\mathcal{L}^2} f,$$

and f is the limit in $\mathcal{L}^2$ of the sequence (f_n).

Example 1.23

(i) In Example 1.11(ii) we saw that, pointwise,

$$x^n \to \begin{cases} 0, & 0 \leq x < 1 \\ 1, & x = 1. \end{cases}$$

Since $\mathcal{L}^2([0, 1]) = \mathcal{L}^2([0, 1))$, we have

$$\| x^n - 0 \| = \left[\int_0^1 x^{2n} dx \right]^{1/2} = \left[\frac{1}{2n + 1} \right]^{1/2} \to 0.$$

Therefore $x^n \xrightarrow{\mathcal{L}^2} 0$.

(ii) The sequence of functions (f_n) defined in Example 1.12 by

$$f_n(x) = \begin{cases} 0, & x = 0 \\ n, & 0 < x \le 1/n \\ 0, & 1/n < x \le 1 \end{cases}$$

also converges pointwise to 0 on $[0, 1]$. But in this case,

$$\| f_n - 0 \|^2 = \int_0^1 f_n^2(x)dx$$

$$= \int_0^{1/n} n^2 dx$$

$$= n \ \text{ for all } n \in \mathbb{N}.$$

Thus $\| f_n - 0 \| = \sqrt{n} \to \infty$, which means the sequence f_n does not converge to 0 in $\mathcal{L}^2$.

This last example shows that pointwise convergence does not imply convergence in $\mathcal{L}^2$. Conversely, convergence in $\mathcal{L}^2$ cannot imply pointwise convergence, since the limit in this case is a class of functions which are equal almost everywhere, not necessarily pointwise. It is legitimate to ask, however, whether a sequence that converges pointwise to some limit f can converge to a different limit in $\mathcal{L}^2$. For example, can the sequence (f_n) in Example 1.23(ii) converge in $\mathcal{L}^2$ to some function other than 0? The answer is no. In other words, if a sequence converges both pointwise and in $\mathcal{L}^2$, then its limit is the same in both cases. More precisely, we should say that the two limits are not distinguishable in $\mathcal{L}^2$ as they are equal almost everywhere.

On the other hand, uniform convergence $f_n \overset{u}{\to} f$ over I implies pointwise convergence, as we have already observed, and we now show that it also implies $f_n \overset{\mathcal{L}^2}{\to} f$ provided the sequence (f_n) and f lie in $\mathcal{L}^2(I)$ and I is bounded: the convergence $(f_n - f) \overset{u}{\to} 0$ implies $|f_n - f|^2 \overset{u}{\to} 0$ (Exercise 1.26). By Theorem 1.14(ii), we therefore have

$$\lim_{n \to \infty} \| f_n - f \|^2 = \lim_{n \to \infty} \int_I |f_n(x) - f(x)|^2 \, dx$$

$$= \int_I \lim_{n \to \infty} |f_n(x) - f(x)|^2 \, dx = 0.$$

The condition that f belong to $\mathcal{L}^2(I)$ is actually not needed, as we shall discover in Theorem 1.26.

Example 1.24 We saw in Example 1.18 that

$$S_n(x) = \sum_{k=1}^{n} \frac{1}{k^2} \sin kx \xrightarrow{u} S(x) = \sum_{k=1}^{\infty} \frac{1}{k^2} \sin kx, \ x \in \mathbb{R},$$

hence the function $S(x)$ is continuous on $[-\pi, \pi]$. Moreover, both S_n and S lie in $\mathcal{L}^2(-\pi, \pi)$ since both are continuous on $[-\pi, \pi]$, so we can write

$$\lim_{n \to \infty} \sum_{k=1}^{n} \frac{1}{k^2} \sin kx = \sum_{k=1}^{\infty} \frac{1}{k^2} \sin kx \text{ in } \mathcal{L}^2(-\pi, \pi).$$

The series $\sum \sin kx / k$, on the other hand, cannot be tested for convergence in $\mathcal{L}^2$ with the tools available, and we have to develop the theory a little further. First we define a *Cauchy sequence* in $\mathcal{L}^2$ along the lines of the corresponding notion in $\mathbb{R}$. This will allow us to test a sequence for convergence without having to guess its limit beforehand.

Definition 1.25 A sequence of functions f_n in $\mathcal{L}^2$ is called a *Cauchy sequence* if, for every $\varepsilon > 0$, there is an integer N such that

$$m, n \geq N \Rightarrow \|f_n - f_m\| < \varepsilon.$$

Clearly, every convergent sequence in $\mathcal{L}^2$ is a Cauchy sequence; for if $f_n \xrightarrow{\mathcal{L}^2} f$, then, by the triangle inequality,

$$\|f_n - f_m\| \leq \|f_n - f\| + \|f_m - f\|,$$

and we can make the right-hand side of this inequality arbitrarily small by taking m and n large enough. The converse of this statement, i.e., that every Cauchy sequence in $\mathcal{L}^2$ converges to some function in $\mathcal{L}^2$, is also true and expresses the *completeness property* of $\mathcal{L}^2$.

Theorem 1.26 (Completeness of $\mathcal{L}^2$) *For every Cauchy sequence (f_n) in $\mathcal{L}^2$ there is a function $f \in \mathcal{L}^2$ such that $f_n \xrightarrow{\mathcal{L}^2} f$.*

There is another theorem which states that, for every function $f \in \mathcal{L}^2(a, b)$, there is a sequence of continuous functions (f_n) on $[a, b]$ such that $f_n \xrightarrow{\mathcal{L}^2} f$. In other words, the set of functions $C([a, b])$ is *dense* in $\mathcal{L}^2(a, b)$ in much the same way that the rationals $\mathbb{Q}$ are dense in $\mathbb{R}$, keeping in mind of course the different topologies of $\mathbb{R}$ and $\mathcal{L}^2$, the first being defined by the absolute value $|\cdot|$ and the second by the norm $\|\cdot\|$. For example, the $\mathcal{L}^2(-1, 1)$ function

$$f(x) = \begin{cases} 0, & -1 \leq x < 0 \\ 1, & 0 \leq x \leq 1, \end{cases}$$

which is discontinuous at $x = 0$, can be approached in the $\mathcal{L}^2$ norm by the sequence of continuous functions

$$f_n = \begin{cases} 0, & -1 \le x \le -1/n \\ nx + 1, & -1/n < x < 0 \\ 1, & 0 \le x \le 1. \end{cases}$$

This is clear from

$$\lim_{n \to \infty} \| f_n - f \| = \lim_{n \to \infty} \left[\int_{-1}^{1} |f_n(x) - f(x)|^2 \, dx \right]^{1/2}$$

$$= \lim_{n \to \infty} \left[\int_{-1/n}^{0} (nx + 1)^2 dx \right]^{1/2}$$

$$= \lim_{n \to \infty} 1/\sqrt{3n} = 0.$$

Needless to say, there are many other sequences in $C([-1, 1])$ which converge to f in $\mathcal{L}^2(-1, 1)$, just as there are many sequences in $\mathbb{Q}$ which converge to the same irrational number.

As we shall have occasion to refer to this result in the following chapter, we give here its precise statement.

Theorem 1.27 (Density of C in $\mathcal{L}^2$) *For any $f \in \mathcal{L}^2(a, b)$ and any $\varepsilon > 0$, there is a continuous function g on $[a, b]$ such that $\| f - g \| < \varepsilon$.*

The proofs of Theorems 1.26 and 1.27 may be found in [15]. The space $\mathcal{L}^2$ is one of the most important examples of a *Hilbert space,* which is an inner product space that is complete under the norm defined by the inner product. It is named after *David Hilbert* (1862–1943), the German mathematician whose work and inspiration did much to develop the ideas of Hilbert space (see [7], vol I). Many of the ideas that we work with will be articulated within the context of $\mathcal{L}^2$.

Example 1.28 Using Theorem 1.26, we can now look into the question of convergence of the sequence $S_n(x) = \sum_{k=1}^{n} \sin kx / k$ in $\mathcal{L}^2(-\pi, \pi)$. Noting that

$$\| S_n(x) - S_m(x) \|^2 = \left\| \sum_{k=m+1}^{n} \frac{1}{k} \sin kx \right\|^2, \quad m < n,$$

we can use the orthogonality of $\{\sin kx : k \in \mathbb{N}\}$ in $\mathcal{L}^2(-\pi, \pi)$ (Example 1.20) to obtain

$$\left\| \sum_{k=m+1}^{n} \frac{1}{k} \sin kx \right\|^2 = \sum_{k=m+1}^{n} \frac{1}{k^2} \|\sin kx\|^2 = \pi \sum_{k=m+1}^{n} \frac{1}{k^2}.$$

Suppose $\varepsilon > 0$. Since $\sum 1/k^2$ is convergent, we can choose N so that

$$n > m \geq N \Rightarrow \sum_{k=m+1}^{n} \frac{1}{k^2} < \frac{\varepsilon^2}{\pi}$$

$$\Rightarrow \|S_n(x) - S_m(x)\| < \varepsilon.$$

Therefore $\sum_{k=1}^{n} \sin kx/k$ is a Cauchy sequence and hence converges in $\mathcal{L}^2(-\pi, \pi)$, though we cannot as yet tell to what limit. Similarly, the series $\sum \cos kx/k$ converges in $\mathcal{L}^2(-\pi, \pi)$, though this series diverges pointwise at all integral multiples of 2π.

This section was devoted to convergence in $\mathcal{L}^2$ because of its importance to the theory of Fourier series, but we could just as easily have been discussing convergence in the weighted space $\mathcal{L}^2_\rho$. Definitions 1.22 and 1.25 and Theorems 1.26 and 1.27 would remain unchanged, with the norm $\|\cdot\|$ replaced by $\|\cdot\|_\rho$ and convergence in $\mathcal{L}^2$ by convergence in $\mathcal{L}^2_\rho$.

Exercises

1.48 Determine the limit in $\mathcal{L}^2$ of each of the following sequences where it exists:

   (a) $f_n(x) = \sqrt[n]{x}, \ 0 \leq x \leq 1,$

   (b) $f_n(x) = \begin{cases} nx, & 0 \leq x < 1/n \\ 1, & 1/n \leq x \leq 1, \end{cases}$

   (c) $f_n(x) = nx(1 - x)^n, \ 0 \leq x \leq 1.$

1.49 Test the following series for convergence in $\mathcal{L}^2\,(-\pi, \pi)$:

   (a) $\displaystyle\sum_{1}^{\infty} \frac{1}{k^{2/3}} \sin kx,$

   (b) $\displaystyle\sum_{1}^{\infty} \frac{1}{k} e^{ikx},$

   (c) $\displaystyle\sum_{1}^{\infty} \frac{1}{\sqrt{k + 1}} \cos kx.$

1.50 If (f_n) is a sequence in $\mathcal{L}^2(a, b)$ which converges to f in $\mathcal{L}^2$, show that $\langle f_n, g \rangle \to \langle f, g \rangle$ for any $g \in \mathcal{L}^2(a, b)$.

1.51 Prove that $|\, \|f\| - \|g\| \,| \leq \|f - g\|$, and hence conclude that if $f_n \overset{\mathcal{L}^2}{\to} f$ then $\|f_n\| \to \|f\|$.

1.52 If the numerical series $\sum |a_n|$ is convergent, prove that $\sum |a_n|^2$ is also convergent, and that the series $\sum a_n \sin nx$ and $\sum a_n \cos nx$ are both continuous on $[-\pi, \pi]$.

1.53 Prove that if the weight functions ρ and σ are related by $\rho \geq \sigma$ on (a, b), then a sequence which converges in $\mathcal{L}^2_\rho(a, b)$ also converges in $\mathcal{L}^2_\sigma(a, b)$.

1.6   Orthogonal Functions

Let

$$\{\varphi_1, \varphi_2, \varphi_3, \cdots\}$$

be an orthogonal set of (non-zero) functions in the complex space $\mathcal{L}^2(I)$, which may be finite or infinite, and suppose that the function $f \in \mathcal{L}^2$ is a finite linear combination of elements in the set $\{\varphi_i\}$,

$$f = \sum_{i=1}^n \alpha_i \varphi_i, \quad \alpha_i \in \mathbb{C}. \tag{1.21}$$

The real interval I plays no role in what follows, so it will not be explicitly mentioned. Taking the inner product of f with φ_k,

$$\langle f, \varphi_k \rangle = \alpha_k \left\| \varphi_k \right\|^2 \text{ for any } k = 1, \cdots, n,$$

we conclude that

$$\alpha_k = \frac{\langle f, \varphi_k \rangle}{\left\| \varphi_k \right\|^2},$$

and the representation (1.21) takes the form

$$f = \sum_{k=1}^n \frac{\langle f, \varphi_k \rangle}{\left\| \varphi_k \right\|^2} \varphi_k. \tag{1.22}$$

In other words, the coefficients α_k in the linear combination (1.21) are determined by the projections of f on φ_k divided by $\left\| \varphi_k \right\|$. In terms of the corresponding orthonormal set $\{\psi_k = \varphi_k / \left\| \varphi_k \right\|\}$,

$$f = \sum_{k=1}^n \langle f, \psi_k \rangle \psi_k,$$

and the coefficients coincide with the projections of f on ψ_k.

Suppose, on the other hand, that f is an arbitrary function in $\mathcal{L}^2$ and that we want to obtain the best approximation of f in $\mathcal{L}^2$, that is, in the norm $\left\| \cdot \right\|$, by a finite linear combination of the elements of $\{\varphi_k\}$. We should then look for the coefficients α_k which minimize the non-negative number

$$\left\| f - \sum_{k=1}^{n} \alpha_k \varphi_k \right\|.$$

We have

$$\left\| f - \sum_{k=1}^{n} \alpha_k \varphi_k \right\|^2 = \left\langle f - \sum_{k=1}^{n} \alpha_k \varphi_k, \ f - \sum_{k=1}^{n} \alpha_k \varphi_k \right\rangle$$

$$= \|f\|^2 - 2 \sum_{k=1}^{n} \mathrm{Re}\,\bar{\alpha}_k \langle f, \varphi_k \rangle + \sum_{k=1}^{n} |\alpha_k|^2 \|\varphi_k\|^2$$

$$= \|f\|^2 - \sum_{k=1}^{n} \frac{|\langle f, \varphi_k \rangle|^2}{\|\varphi_k\|^2}$$

$$+ \sum_{k=1}^{n} \|\varphi_k\|^2 \left[\left| \alpha_k^2 \right| - 2\mathrm{Re}\,\bar{\alpha}_k \frac{\langle f, \varphi_k \rangle}{\|\varphi_k\|^2} + \frac{|\langle f, \varphi_k \rangle|^2}{\|\varphi_k\|^4} \right]$$

$$= \|f\|^2 - \sum_{k=1}^{n} \frac{|\langle f, \varphi_k \rangle|^2}{\|\varphi_k\|^2} + \sum_{k=1}^{n} \|\varphi_k\|^2 \left| \alpha_k - \frac{\langle f, \varphi_k \rangle}{\|\varphi_k\|^2} \right|^2.$$

Since the coefficients α_k appear only in the last term

$$\sum_{k=1}^{n} \|\varphi_k\|^2 \left| \alpha_k - \frac{\langle f, \varphi_k \rangle}{\|\varphi_k\|^2} \right|^2 \geq 0,$$

we obviously achieve the minimum of $\left\| f - \sum_{k=1}^{n} \alpha_k \varphi_k \right\|^2$, and hence of $\left\| f - \sum_{k=1}^{n} \alpha_k \varphi_k \right\|$, by choosing

$$\alpha_k = \frac{\langle f, \varphi_k \rangle}{\|\varphi_k\|^2}.$$

This minimum is given by

$$\left\| f - \sum_{k=1}^{n} \frac{\langle f, \varphi_k \rangle}{\|\varphi_k\|^2} \varphi_k \right\|^2 = \|f\|^2 - \sum_{k=1}^{n} \frac{|\langle f, \varphi_k \rangle|^2}{\|\varphi_k\|^2} \geq 0, \tag{1.23}$$

and yields the relation,

$$\sum_{k=1}^{n} \frac{\left|\langle f, \varphi_k \rangle\right|^2}{\left\|\varphi_k\right\|^2} \leq \|f\|^2 .$$

Since this relation is true for any n, it is also true in the limit as $n \to \infty$. The resulting inequality

$$\sum_{k=1}^{\infty} \frac{\left|\langle f, \varphi_k \rangle\right|^2}{\left\|\varphi_k\right\|^2} \leq \|f\|^2 , \tag{1.24}$$

known as *Bessel's inequality*, holds for any orthogonal set $\{\varphi_k : k \in \mathbb{N}\}$ and any f in $\mathcal{L}^2$.

In view of (1.23), Bessel's inequality becomes an equality if, and only if,

$$\left\| f - \sum_{k=1}^{\infty} \frac{\langle f, \varphi_k \rangle}{\left\|\varphi_k\right\|^2} \varphi_k \right\| = 0,$$

or, equivalently,

$$f = \sum_{k=1}^{\infty} \frac{\langle f, \varphi_k \rangle}{\left\|\varphi_k\right\|^2} \varphi_k \ \text{ in } \mathcal{L}^2,$$

which means that f is represented in $\mathcal{L}^2$ by the sum $\sum_{k=1}^{\infty} \alpha_k \varphi_k$, where $\alpha_k = \langle f, \varphi_k \rangle / \left\|\varphi_k\right\|^2$.

Definition 1.29 An orthogonal set $\{\varphi_n : n \in \mathbb{N}\}$ in $\mathcal{L}^2$ is said to be *complete* if, for any $f \in \mathcal{L}^2$,

$$\sum_{k=1}^{n} \frac{\langle f, \varphi_k \rangle}{\left\|\varphi_k\right\|^2} \varphi_k \xrightarrow{\mathcal{L}^2} f.$$

Thus a complete orthogonal set in $\mathcal{L}^2$ becomes a *basis* for the space, and since $\mathcal{L}^2$ is infinite dimensional the basis has to be an infinite set. When Bessel's inequality becomes an equality, the resulting relation

$$\|f\|^2 = \sum_{n=1}^{\infty} \frac{\left|\langle f, \varphi_n \rangle\right|^2}{\left\|\varphi_n\right\|^2} \tag{1.25}$$

is called *Parseval's relation* or the *completeness relation*. The second term is justified by the following theorem, which is really a restatement of Definition 1.29.

Theorem 1.30 *An orthogonal set* $\{\varphi_n : n \in \mathbb{N}\}$ *is complete if, and only if, it satisfies Parseval's relation* (1.25) *for any* $f \in \mathcal{L}^2$.

Remark 1.31

1. Given any orthogonal set $\{\varphi_n : n \in \mathbb{N}\}$ in $\mathcal{L}^2$, we have shown that we obtain the best $\mathcal{L}^2$-approximation

$$\sum_{k=1}^{n} \alpha_k \varphi_k$$

   of the function $f \in \mathcal{L}^2$ by choosing $\alpha_k = \langle f, \varphi_k \rangle / \|\varphi_k\|^2$, and this choice is independent of n. If $\{\varphi_n\}$ is complete then the equality $f = \sum_{n=1}^{\infty} \alpha_n \varphi_n$ holds in $\mathcal{L}^2$.

2. When the orthogonal set $\{\varphi_n\}$ is normalized to $\{\psi_k = \varphi_k / \|\varphi_k\|\}$, Bessel's inequality takes the form

$$\sum_{k=1}^{\infty} |\langle f, \psi_k \rangle|^2 \le \|f\|^2,$$

   and Parseval's relation becomes

$$\|f\|^2 = \sum_{n=1}^{\infty} |\langle f, \psi_n \rangle|^2.$$

3. For any $f \in \mathcal{L}^2$, since $\|f\| < \infty$, we conclude from Bessel's inequality that $\langle f, \psi_n \rangle \to 0$ whether the orthonormal set $\{\psi_n\}$ is complete or not.

Parseval's relation may be regarded as a generalization of the theorem of Pythagoras from $\mathbb{R}^n$ to $\mathcal{L}^2$, where $\|f\|^2$ corresponds to the square of the length of the vector, and $\sum_{n=1}^{\infty} |\langle f, \psi_n \rangle|^2$ represents the sum of the squares of its projections on the orthonormal basis. That is one reason why $\mathcal{L}^2$ is considered the natural generalization of the finite-dimensional Euclidean space to infinite dimensions. It preserves some of the basic geometric structure of $\mathbb{R}^n$ in some form, and the completeness property (Theorem 1.26) guarantees its closure under limiting operations on Cauchy sequences.

Exercises

1.54 Use the the familiar trigonometric identities to express $\sin^3 x$ as a linear combination of the orthogonal functions

$$\{1, \cos x, \sin x, \cos 2x, \sin 2x, \cos 3x, \sin 3x\}$$

for all $x \in [-\pi, \pi]$.

1.55 Check the result of Exercise 1.54 against Eq. (1.22) by calculating the coefficient $\langle \sin^3 x, \sin x \rangle / \|\sin x\|^2$ in the orthogonal expansion of $\sin^3 x$.

1.56 If l is any positive number, show that $\{\sin(n\pi x/l) : n \in \mathbb{N}\}$ and $\{\cos(n\pi x/l) : n \in \mathbb{N}_0\}$ are orthogonal sets in $\mathcal{L}^2(0, l)$. Determine the corresponding orthonormal sets.

1.57 Determine the coefficients c_i in the linear combination

$$c_1 + c_2 \sin \pi x + c_3 \sin 2\pi x$$

which give the best approximation in $\mathcal{L}^2(0, 2)$ of the function $f(x) = x$, $0 < x < 2$.

1.58 Determine the coefficients a_i and b_i in the linear combination

$$a_0 + a_1 \cos x + b_1 \sin x + a_2 \cos 2x + b_2 \sin 2x$$

which give the best approximation in $\mathcal{L}^2(-\pi, \pi)$ of $f(x) = |x|$, $-\pi \le x \le \pi$.

1.59 Let $p_0, p_1,$ and p_2 be the three orthogonal polynomials formed from the set $\{1, x, x^2\}$ by the Gram-Schmidt method, where $-1 \le x \le 1$. Determine the constant coefficients in the second-degree polynomial $a_0 p_0(x) + a_1 p_1(x) + a_2 p_2(x)$ which give the best approximation in $\mathcal{L}^2(-1, 1)$ of e^x. Can you think of another polynomial p of degree 2 which approximates e^x on $(-1, 1)$ in a different sense?

1.60 Assuming that

$$1 - x = \frac{8}{\pi^2} \sum_{n=1}^{\infty} \frac{1}{(2n-1)^2} \cos \frac{(2n-1)\pi}{2} x, \quad 0 \le x \le 2,$$

use Parseval's identity to prove that

$$\pi^4 = 96 \sum_{n=1}^{\infty} \frac{1}{(2n-1)^4}.$$

1.61 Define a real sequence (a_k) such that $\sum a_k^2$ converges and $\sum a_k$ diverges. What type of convergence can the series $\sum a_n \cos nx$ have on $[-\pi, \pi]$?

1.62 Suppose $\{f_n : n \in \mathbb{N}\}$ is an orthogonal set in $\mathcal{L}^2(0, l)$, and let

$$\varphi_n(x) = \left\{ \begin{array}{l} f_n(x), \ x \in [0, l] \\ f_n(-x), \ x \in [-l, 0) \end{array} \right\}$$

$$\psi_n(x) = \left\{ \begin{array}{l} f_n(x), \ x \in [0, l] \\ -f_n(-x), \ x \in [-l, 0) \end{array} \right\}$$

be the even and odd extensions, respectively, of f_n from $[0, l]$ to $[-l, l]$. Show that the set $\{\varphi_n\} \cup \{\psi_n\}$ is orthogonal in $\mathcal{L}^2(-l, l)$. If $\{f_n\}$ is orthonormal in $\mathcal{L}^2(0, l)$, what is the corresponding orthonormal set in $\mathcal{L}^2(-l, l)$?

Chapter 2
The Sturm-Liouville Theory

Complete orthogonal sets of functions in $\mathcal{L}^2$ which span the space arise naturally as solutions of certain second-order linear differential equations under appropriate boundary conditions, commonly referred to as *Sturm-Liouville boundary-value problems*, after the Swiss mathematician *Jacques Sturm* (1803–1855) and the French mathematician *Joseph Liouville* (1809–1882), who studied these problems and the properties of their solutions. The differential equations considered here arise directly as mathematical models of physical phenomena, but more often as a result of using the method of separation of variables to solve the classical partial differential equations of physics. In this chapter the linear second order differential equation is posed as an eigenvalue problem for a self-adjoint differential operator, and the completeness of its eigenfunctions in $\mathcal{L}^2$ is proved for the regular Sturm-Liouville problem.

2.1 Linear Second Order Equations

Consider the linear ordinary differential equation of second order on the real interval I given by

$$a_0(x)y'' + a_1(x)y' + a_2(x)y = f(x), \tag{2.1}$$

where the coefficients a_0, a_1, a_2, and f are given complex functions on I. When $f = 0$ on I, the equation is called *homogeneous*, otherwise it is *non-homogeneous*. Any (complex) function $\varphi \in C^2(I)$ is a *solution* of Eq. (2.1) if the substitution of y by φ results in the identity

$$a_0(x)\varphi''(x) + a_1(x)\varphi'(x) + a_2(x)\varphi(x) = f(x) \ \text{ for all } x \in I.$$

© The Author(s), under exclusive license to Springer-Verlag London Ltd.,
part of Springer Nature 2026
M. A. Al-Gwaiz, *Sturm-Liouville Theory and its Applications*, Springer
Undergraduate Mathematics Series, https://doi.org/10.1007/978-1-4471-7610-7_2

If we denote the linear second order differential operator

$$a_0(x)\frac{d^2}{dx^2} + a_1(x)\frac{d}{dx} + a_2(x)$$

by L, then Eq. (2.1) can be written in the form $Ly = f$. The differential operator L is linear, in the sense that

$$L(c_1\varphi + c_2\psi) = c_1 L\varphi + c_2 L\psi$$

for any functions $\varphi, \psi \in C^2(I)$ and any constants $c_1, c_2 \in \mathbb{C}$. Unless otherwise specified, all differential equations and operators that we deal with are linear. A fundamental property of a linear homogeneous differential equation is that any linear combination of solutions of the equation is also a solution; for if φ and ψ satisfy

$$L\varphi = 0, \ \ L\psi = 0,$$

then we clearly have $L(c_1\varphi + c_2\psi) = 0$ for any pair of constants c_1 and c_2. This is known as the *superposition principle*.

If the function a_0 does not vanish at any point on I, Eq. (2.1) may be divided by a_0 to give

$$y'' + q(x)y' + r(x)y = g(x), \tag{2.2}$$

where $q = a_1/a_0$, $r = a_2/a_0$, and $g = f/a_0$. Eqs. (2.1) and (2.2) are clearly equivalent, in the sense that they have the same set of solutions. Equation (2.1) is then said to be *regular* on I; otherwise, if there is a point $c \in I$ where $a_0(c) = 0$, the equation is *singular*, and c is referred to as a *singular point* of the equation.

According to the existence and uniqueness theorem for linear equations (see [6]), if the functions q, r, and g are all continuous on I and x_0 is a point in I, then, for any two numbers ξ and η, there is a unique solution φ of (2.2) on I such that

$$\varphi(x_0) = \xi, \ \ \varphi'(x_0) = \eta. \tag{2.3}$$

Equations (2.3) are called *initial conditions*, and the system of Eqs. (2.2) and (2.3) is called an *initial-value problem*.

Here we list some well-known properties of the solutions of Eq. (2.2), which may be found in many standard introductions to ordinary differential equations:

1. The homogeneous equation

$$y'' + q(x)y' + r(x)y = 0, \ \ x \in I, \tag{2.4}$$

has two linearly independent solutions $y_1(x)$ and $y_2(x)$ on I. A linear combination of these two solutions

$$c_1 y_1 + c_2 y_2, \tag{2.5}$$

where c_1 and c_2 are arbitrary constants, is the *general solution* of (2.4); that is, any solution of the equation is given by (2.5) for some values of c_1 and c_2. These constants are determined uniquely by the initial conditions (2.3). When $c_1 = c_2 = 0$ we obtain the so-called *trivial solution* $y \equiv 0$, which is always a solution of the homogeneous equation. By the uniqueness theorem, it is the only solution if $\xi = \eta = 0$ in (2.3).

2. If $y_p(x)$ is any particular solution of the non-homogeneous Eq. (2.2), then

$$y_p + c_1 y_1 + c_2 y_2$$

is the general solution of (2.2). By applying the initial conditions (2.3), the constants c_1 and c_2 are determined and we obtain the unique solution of the system of Eqs. (2.2) and (2.3).

3. When the coefficients q and r are constants, the general solution of Eq. (2.4) has the form

$$c_1 e^{m_1 x} + c_2 e^{m_2 x},$$

where m_1 and m_2 are the roots of the second degree equation $m^2 + qm + r = 0$ when the roots are distinct. If $m_1 = m_2 = m$, then the solution takes the form $c_1 e^{mx} + c_2 x e^{mx}$, in which the functions e^{mx} and $x e^{mx}$ are clearly linearly independent.

4. When $a_0(x) = x^2$, $a_1(x) = ax$, and $a_2(x) = b$, where a and b are constants, the homogeneous version of Eq. (2.1) becomes

$$x^2 y'' + axy' + by = 0,$$

which is called the *Cauchy-Euler equation.* On $(0, \infty)$ its general solution is given by

$$c_1 x^{m_1} + c_2 x^{m_2},$$

where m_1 and m_2 are the distinct roots of $m^2 + (a - 1)m + b = 0$. When $m_1 = m_2 = m$, the general solution is $c_1 x^m + c_2 x^m \log x$.

5. If the coefficients q and r are analytic functions at some point x_0 in the interior of I, which means each may be represented in an open interval centered at x_0 by a power series in $(x - x_0)$, then the general solution of (2.4) is also analytic at

x_0, and is represented by a power series of the form

$$\sum_{n=0}^{\infty} c_n (x - x_0)^n .$$

The series converges in the intersection of the two intervals of convergence (of q and r) and I. Substituting this series into Eq. (2.4) allows us to express the coefficients c_n for all $n \in \{2, 3, 4, \cdots\}$ in terms of c_0 and c_1, which remain arbitrary.

With $I = [a, b]$, the solutions of Eq. (2.1) may be subjected to two *boundary conditions*. These can take one of the following forms:

$$(i) \;\; y(c) = \xi, \;\; y'(c) = \eta, \;\; c \in \{a, b\},$$

$$(ii) \;\; y(a) = \xi, \;\; y(b) = \eta,$$

$$(iii) \;\; y'(a) = \xi, \;\; y'(b) = \eta.$$

When the boundary conditions are given at the same point c, as in (i), they are referred to as initial conditions as mentioned earlier, especially when time is the independent variable. For the purpose of obtaining a unique solution of Eq. (2.1), the point c need not, in general, be one of the end-points of the interval I, and can be any interior point. But in this presentation, as in most physical applications, boundary (or initial) conditions will always be imposed at the end-points of I. The forms (i) to (iii) of the boundary conditions may be expressed by the pair of equations

$$\alpha_1 y(a) + \alpha_2 y'(a) + \alpha_3 y(b) + \alpha_4 y'(b) = \xi, \tag{2.6}$$

$$\beta_1 y(b) + \beta_2 y'(b) + \beta_3 y(a) + \beta_4 y'(a) = \eta, \tag{2.7}$$

where α_i and β_i are constants that satisfy $\sum_{i=1}^{4} |\alpha_i| > 0$ and $\sum_{i=1}^{4} |\beta_i| > 0$, that is, such that not all the α_i or β_i are zeros. The system of Eqs. (2.1), (2.6), and (2.7) is called a *boundary-value problem*.

The boundary conditions (2.6) and (2.7) are called *homogeneous* if $\xi = \eta = 0$, and *separated* if $\alpha_3 = \alpha_4 = \beta_3 = \beta_4 = 0$. Separated boundary conditions, which have the form

$$\alpha_1 y(a) + \alpha_2 y'(a) = \xi, \;\; \beta_1 y(b) + \beta_2 y'(b) = \eta, \tag{2.8}$$

are of particular significance in this study. Another important pair of homogeneous conditions, which result from a special choice of the coefficients in (2.6) and (2.7), is given by

$$y(a) = y(b), \;\; y'(a) = y'(b). \tag{2.9}$$

Equation (2.9) are called *periodic boundary conditions*. It should be noted that periodic conditions are coupled, not separated.

Definition 2.1 For any two functions $f, g \in C^1$ the determinant

$$W(f, g)(x) = \begin{vmatrix} f(x) & g(x) \\ f'(x) & g'(x) \end{vmatrix} = f(x)g'(x) - g(x)f'(x)$$

is called the *Wronskian* of f and g, and is denoted by the symbol $W(f, g)(x)$ or $W(x)$.

The Wronskian derives its significance in this study from the following lemmas.

Lemma 2.2 *If y_1 and y_2 are solutions of the homogeneous equation*

$$y'' + q(x)y' + r(x)y = 0, \quad x \in I, \tag{2.10}$$

where $q \in C(I)$, then either $W(y_1, y_2)(x) = 0$ for all $x \in I$, or $W(y_1, y_2)(x) \neq 0$ for any $x \in I$.

Proof From Definition 2.1,

$$W' = y_1 y_2'' - y_2 y_1''.$$

Since y_1 and y_2 are solutions of Eq. (2.10), we have

$$y_1'' + q y_1' + r y_1 = 0,$$
$$y_2'' + q y_2' + r y_2 = 0.$$

Multiplying the first equation by y_2, the second by y_1, and subtracting, yields

$$y_1 y_2'' - y_2 y_1'' + q(y_1 y_2' - y_2 y_1') = 0$$
$$\Rightarrow \quad W' + qW = 0.$$

Integrating this last equation, we obtain

$$W(x) = c \exp\left(-\int_a^x q(t)dt\right), \quad x \in I, \tag{2.11}$$

where c is an arbitrary constant. The exponential function does not vanish for any (real or complex) exponent, therefore $W(x) = 0$ if, and only if, $c = 0$. $\qquad\square$

Remark 2.3 With $q \in C(I)$, both W and W' in the expression (2.11) are continuous.

Lemma 2.4 *Any two solutions y_1 and y_2 of Eq. (2.10) are linearly independent if, and only if, $W(y_1, y_2)(x) \neq 0$ on I.*

Proof If y_1 and y_2 are linearly dependent, then one of them is a constant multiple of the other, and therefore $W(y_1, y_2)(x) = 0$ on I. Conversely, if $W(y_1, y_2)(x) = 0$ at any point in I, then, by Lemma 2.2, $W(y_1, y_2) \equiv 0$. From the properties of determinants, this implies that y_1 and y_2 are linearly dependent [6].   □

Remark 2.5 We used the fact that y_1 and y_2 are solutions of Eq. (2.10) only in the second part of the proof, the "only if" part. That is because the Wronskian of two linearly independent functions (which are not solutions) may vanish at some, but not all, points in I. Consider, for example, x and x^2 on $[-1, 1]$.

Example 2.6 The equation

$$y'' + y = 0 \tag{2.12}$$

has the two linearly independent solutions, $\sin x$ and $\cos x$. Hence the general solution is

$$y(x) = c_1 \cos x + c_2 \sin x.$$

Note that

$$W(\cos x, \sin x) = \cos^2 x + \sin^2 x = 1 \quad \text{for all } x \in \mathbb{R}.$$

If Eq. (2.12) is given on the interval $[0, \pi]$ subject to the initial conditions

$$y(0) = 0, \ y'(0) = 1,$$

we obtain the unique solution

$$y(x) = \sin x.$$

But the homogeneous boundary conditions

$$y(0) = 0, \ y'(0) = 0,$$

yield the trivial solution $y = 0$, as would be expected.
  On the other hand, the boundary conditions

$$y(0) = 0, \ y(\pi) = 0,$$

do not give a unique solution because the pair of equations

$$y(0) = c_1 \cos 0 + c_2 \sin 0 = c_1 = 0$$

$$y(\pi) = c_1 \cos \pi + c_2 \sin \pi = -c_1 = 0$$

do not determine the constant c_2. The determinant of the coefficients in this system is

$$\begin{vmatrix} \cos 0 & \sin 0 \\ \cos \pi & \sin \pi \end{vmatrix} = 0.$$

This last example indicates that the boundary conditions (2.6) and (2.7) do not uniquely determine the constants c_1 and c_2 in the general solution in all cases. But the (initial) conditions

$$y(x_0) = \xi, \quad y'(x_0) = \eta,$$

always give a unique solution, because the determinant of the coefficients in the system of equations

$$c_1 y_1(x_0) + c_2 y_2(x_0) = \xi$$
$$c_1 y_2'(x_0) + c_2 y_2'(x_0) = \eta$$

is given by

$$\begin{vmatrix} y_1(x_0) & y_2(x_0) \\ y_1'(x_0) & y_2'(x_0) \end{vmatrix} = W(y_1, y_2)(x_0),$$

which cannot vanish according to Lemma 2.4 if y_1 and y_2 are linearly independent.

In general, given a second order differential equation on $[a, b]$, the separated boundary conditions

$$\alpha_1 y(a) + \alpha_2 y'(a) = \xi,$$
$$\beta_1 y(b) + \beta_2 y'(b) = \eta,$$

imposed on the general solution $y = c_1 y_1 + c_2 y_2$, yield the system

$$c_1[\alpha_1 y_1(a) + \alpha_2 y_1'(a)] + c_2[\alpha_1 y_2(a) + \alpha_2 y_2'(a)] = \xi,$$
$$c_1[\beta_1 y_1(b) + \beta_2 y_1'(b)] + c_2[\beta_1 y_2(b) + \beta_2 y_2'(b)] = \eta.$$

Hence the constants c_1 and c_2 are uniquely determined if, and only if,

$$\begin{vmatrix} (\alpha_1 y_1 + \alpha_2 y_1')(a) & (\alpha_1 y_2 + \alpha_2 y_2')(a) \\ (\beta_1 y_1 + \beta_2 y_1')(b) & (\beta_1 y_2 + \beta_2 y_2')(b) \end{vmatrix} \neq 0. \tag{2.13}$$

Exercises

2.1 Find the general solution of each of the following equations:

(a) $y'' - 4y' + 7y = e^x$,
(b) $xy'' - y' = 3x^2$,
(c) $x^2 y'' + 3xy' + y = x - 1$.

2.2 Use the transformation $t = \sqrt{x}$ to solve $xy'' + y'/2 - y = 0$.

2.3 Use power series to solve the equation $y'' + 2xy' + 4y = 0$ about the point $x = 0$. Determine the interval of convergence of the solution.

2.4 Solve the initial-value problem

$$y'' + \frac{1}{1-x}(xy' - y) = 0, \quad -1 < x < 1,$$

$$y(0) = 0, \quad y'(0) = 1.$$

2.5 If φ_1, φ_2, and φ_3 are solutions of Eq. (2.4), prove that

$$\begin{vmatrix} \varphi_1 & \varphi_2 & \varphi_3 \\ \varphi_1' & \varphi_2' & \varphi_3' \\ \varphi_1'' & \varphi_2'' & \varphi_3'' \end{vmatrix} = 0.$$

2.6 If y_1 and y_2 are linearly independent solutions of Eq. (2.4), show that

$$q = \frac{y_2 y_1'' - y_1 y_2''}{W(y_1, y_2)}, \quad r = \frac{y_1' y_2'' - y_2' y_1''}{W(y_1, y_2)}.$$

2.7 For each of the following pair of solutions, determine the corresponding differential equation of the form $a_0 y'' + a_1 y' + a_2 y = 0$:

(a) $e^{-x} \cos 2x$, $e^{-x} \sin 2x$,
(b) x, x^{-1},
(c) $1, \log x$.

2.2 Zeros of Solutions

It is not necessary, nor is it always possible, to solve a differential equation of the type

$$y'' + q(x)y' + r(x)y = 0 \tag{2.14}$$

explicitly in order to study the basic properties of its solutions. Under certain conditions, the parameters of the equation and its boundary conditions determine these properties almost completely. In particular, such qualitative features of a solution as the number and distribution of its zeros, its singular points, its asymptotic behaviour, and its orthogonality properties, are all governed by the coefficients q and r and the given boundary conditions. We can therefore attempt to deduce some of these properties by analysing the effect of these coefficients on the behaviour of the solution. In this section we shall investigate the effect of q and r on the distribution of the zeros of the solutions of Eq. (2.14). Orthogonality will be addressed in the next two sections.

In Example 2.6 we found that the two solutions of $y'' + y = 0$ were $\sin x$ and $\cos x$. These have an infinite sequence of alternating zeros distributed uniformly on $\mathbb{R}$, given by

$$\cdots < -\pi < -\frac{\pi}{2} < 0 < \frac{\pi}{2} < \pi < \frac{3\pi}{2} < \cdots,$$

where $\{n\pi : n \in \mathbb{Z}\}$ are the zeros of $\sin x$ and $\{\pi/2 + n\pi : n \in \mathbb{Z}\}$ are those of $\cos x$. We shall presently find that this is not completely accidental.

A function f defined on a real interval I is said to have an *isolated zero* at $x_0 \in I$ if $f(x_0) = 0$ and there is a neighborhood U of x_0 such that $f(x) \neq 0$ for all $x \in I \cap U \backslash \{x_0\}$.

In this section the functions $q,\ r \in C(I)$ in Eq. (2.14) as well as the solution $y \in C^2(I)$ are all assumed to be real, though the proofs below only require the continuity of y'.

Lemma 2.7 *If y is a non-trivial solution of the homogeneous Eq. (2.14) on the interval I, then the zeros of y are isolated in I.*

Proof Suppose $y(x_0) = 0$, where y is a solution of (2.14). If $y'(x_0) = 0$, then y is identically 0 by the uniqueness theorem for second order differential equations. If $y'(x_0) \neq 0$ then, since y' is continuous on I, there is a neighborhood U of x_0 where $y' \neq 0$ on $U \cap I$ (see [1]). Consequently y is either strictly increasing or strictly decreasing on $U \cap I$.   □

Theorem 2.8 (Sturm Separation Theorem) *If y_1 and y_2 are linearly independent solutions of the equation*

$$y'' + q(x)y' + r(x)y = 0, \quad x \in I,$$

then the zeros of y_1 are distinct from those of y_2, and the two sequences of zeros alternate, that is, y_1 has exactly one zero between any two successive zeros of y_2, and vice-versa (Fig. 2.1).

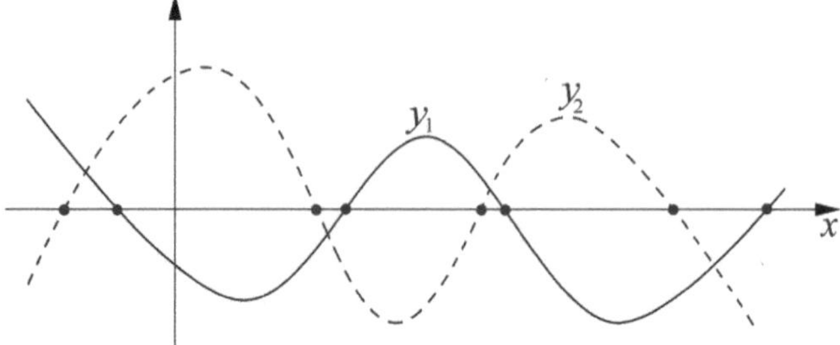

Fig. 2.1 The alternating zeros of y_1 and y_2

Proof Since y_1 and y_2 are linearly independent, their Wronskian

$$W(y_1, y_2)(x) = y_1(x)y_2'(x) - y_2(x)y_1'(x)$$

does not vanish and has therefore one sign on I (Lemma 2.4). Note first that y_1 and y_2 cannot have a common zero, otherwise W would vanish there. Suppose x_1 and x_2 are two successive zeros of y_2. Then

$$W(x_1) = y_1(x_1)y_2'(x_1) \neq 0,$$
$$W(x_2) = y_1(x_2)y_2'(x_2) \neq 0,$$

and the numbers $y_1(x_1)$, $y_1(x_2)$, $y_2'(x_1)$, and $y_2'(x_2)$ are all non-zero. Since y_2' is continuous on I, x_1 has a neighborhood U_1 where the sign of y_2' does not change, and similarly there is a neighborhood U_2 of x_2 where y_2' does not change sign. But the signs of y_2' in $U_1 \cap I$ and $U_2 \cap I$ cannot be the same, for if y_2 is increasing on one then it has to be decreasing on the other. For $W(x)$ to have a constant sign on I, $y_1(x_1)$ and $y_1(x_2)$ must therefore have opposite signs; hence y_1, being continuous, has at least one zero between x_1 and x_2. There cannot be more than one such zero, for if x_3 and x_4 are two zeros of y_1 which lie between x_1 and x_2, we can use the same argument to conclude that y_2 vanishes between x_3 and x_4. But this contradicts the assumption that x_1 and x_2 are consecutive zeros of y_2. $\square$

Corollary 2.9 *If two solutions of $y'' + q(x)y' + r(x)y = 0$ have a common zero in I, then they are linearly dependent.*

In order to study the distribution of zeros of Eq. (2.14), it would be much more convenient if we could suppress the middle term qy' by transforming the equation to

$$u'' + \rho(x)u = 0. \tag{2.15}$$

To that end, we set

$$y(x) = u(x)v(x),$$

so that

$$y'(x) = u'(x)v(x) + u(x)v'(x),$$
$$y''(x) = u''(x)v(x) + 2u'(x)v'(x) + u(x)v''(x).$$

Substituting into Eq. (2.14) yields

$$vu'' + (2v' + qv)u' + (v'' + qv' + rv)u = 0.$$

Thus we obtain (2.15) by choosing $2v' + qv = 0$, which implies

$$v(x) = \exp\left(-\frac{1}{2}\int_a^x q(t)dt\right), \tag{2.16}$$

$$\rho(x) = r(x) - \frac{1}{4}q^2(x) - \frac{1}{2}q'(x),$$

assuming q is differentiable. Since the exponential function v never vanishes on $\mathbb{R}$, the zeros of u coincide with those of y, and we may, for the purpose of investigating the distribution of zeros of Eq. (2.14), confine our attention to Eq. (2.15).

Theorem 2.10 (Sturm Comparison Theorem) *Let φ and ψ be non-trivial solutions of the equations*

$$y'' + r_1(x)y = 0,$$
$$u'' + r_2(x)u = 0, \; x \in I,$$

respectively, and suppose $r_1(x) \geq r_2(x)$ for all $x \in I$. Then φ has at least one zero between every two consecutive zeros of ψ, unless $r_1(x) \equiv r_2(x)$ and φ is a constant multiple of ψ.

Proof Let x_1 and x_2 be any two consecutive zeros of ψ on I, and suppose that φ has no zero in the open interval (x_1, x_2). Assume that both φ and ψ are positive on (x_1, x_2), otherwise change the sign of the negative function. Since φ' and ψ' are continuous, it follows that $\psi'(x_1) \geq 0$ and $\psi'(x_2) \leq 0$, and therefore the Wronskian of φ and ψ satisfies

$$W(x_1) = \varphi(x_1)\psi'(x_1) \geq 0, \; W(x_2) = \varphi(x_2)\psi'(x_2) \leq 0. \tag{2.17}$$

But since

$$W'(x) = \varphi(x)\psi''(x) - \varphi''(x)\psi(x)$$
$$= [r_1(x) - r_2(x)]\varphi(x)\psi(x) \geq 0 \ \text{ for all } x \in (x_1, x_2),$$

W is an increasing function on (x_1, x_2). This contradicts Eq. (2.17) unless $r_1(x) - r_2(x) \equiv 0$ and $W(x) \equiv 0$, in which case φ and ψ are linearly dependent (by Lemma 2.4). $\qquad\square$

Corollary 2.11 *Let φ be a non-trivial solution of $y'' + r(x)y = 0$ on I. If $r(x) \leq 0$ then φ has at most one zero on I.*

Proof If the solution φ has two zeros on I, say x_1 and x_2, then, by Theorem 2.10, the solution $\psi(x) \equiv 1$ of $u'' = 0$ must vanish on (x_1, x_2), which is impossible. $\quad\square$

Example 2.12

(i) Any non-trivial solution of $y'' = 0$ on $\mathbb{R}$ is a special case of

$$\varphi(x) = c_1 x + c_2,$$

which is represented by a straight line, and has at most one isolated zero.

(ii) The equation $y'' - y = 0$ on $\mathbb{R}$ has the general solution

$$\varphi(x) = c_1 e^x + c_2 e^{-x}, \quad x \in \mathbb{R},$$

which has no zeros unless $c_1 c_2 < 0$, in which case $\varphi(x)$ has one zero at $x = \frac{1}{2}\log(-c_2/c_1)$.

(iii) The solution of $y'' + y = 0$ in $\mathbb{R}$ is given by

$$\varphi(x) = c_1 \cos x + c_2 \sin x = a \sin(x - b),$$

where $a = \sqrt{c_1^2 + c_2^2} \neq 0$ and $b = -\arctan(c_1/c_2)$. φ has an infinite number of zeros in $\mathbb{R}$ given by $x_n = b + n\pi$, $n \in \mathbb{Z}$.

A nontrivial solution of

$$y'' + r(x)y = 0, \ \ x \in I, \tag{2.18}$$

is called *oscillatory* if it has an infinite number of zeros, as in Example 2.12(iii). According to Theorem 2.10, whether this equation has oscillatory solutions or not depends on the function r and the interval I. If $r(x) \leq 0$, the solutions cannot oscillate by Corollary 2.11; but if

$$r(x) > k^2 > 0,$$

for some positive constant k, then any solution of (2.18) on an unbounded interval I has an infinite number of zeros distributed between the zeros of any solution of $y'' + k^2 y = 0$, such as $a \sin k(x - b)$, which are given by

$$x_n = b + \frac{n\pi}{k}.$$

Thus every subinterval of $\mathbb{R}$ of length π/k has at least one zero of Eq. (2.18), and as k increases we would expect the number of zeros to increase. This, of course, is clearly the case when r is constant.

From the Sturm separation theorem we also conclude that, if the interval I is unbounded and one solution of the equation

$$y'' + q(x)y' + r(x)y = 0$$

oscillates, then all its solutions oscillate on I.

Example 2.13 The equation

$$y'' + \frac{1}{x}y' + \left(1 - \frac{v^2}{x^2}\right)y = 0, \quad x \in (0, \infty), \tag{2.19}$$

is known as *Bessel's equation* of order v, and its solutions are the subject of Chap. 5. Using formula (2.16) to define $u = \sqrt{x}\,y$, Bessel's equation under this transformation takes the form

$$u'' + \left(1 + \frac{1 - 4v^2}{4x^2}\right)u = 0. \tag{2.20}$$

By comparing Eq. (2.20) and $u'' + u = 0$, whose general solution is $a \sin (x - b)$, we see that

$$r(x) = 1 + \frac{1 - 4v^2}{4x^2} \begin{cases} \geq 1 \text{ if } 0 \leq v \leq 1/2 \\ < 1 \text{ if } v > 1/2. \end{cases}$$

Applying Theorem 2.10 we therefore conclude:

(i) If $0 \leq v \leq 1/2$ then, in every subinterval of $(0, \infty)$ of length π, any solution of Bessel's equation has at least one zero.

(ii) If $v > 1/2$ then, in every subinterval of $(0, \infty)$ of length π, any nontrivial solution of Bessel's equation has at most one zero.

(iii) If $v = 1/2$, the distance between successive zeros of any nontrivial solution of Bessel's equation is exactly π.

Exercises

2.8   Prove that any nontrivial solution of $y'' + r(x)y = 0$ on a compact interval has
at most a finite number of zeros.

2.9   Prove that any nontrivial solution of the equation

$$y'' + \frac{k}{x^2}y = 0$$

on $(0, \infty)$ is oscillatory if, and only if, $k > 1/4$.

2.10   Use the result of Exercise 2.9 to conclude that any nontrivial solution of the
equation $y'' + r(x)y = 0$ on $(0, \infty)$ has an infinite number of zeros if $r(x) \geq$
k/x^2 for some $k > 1/4$, and only a finite number if $r(x) < 1/4x^2$.

2.11   Let φ be a nontrivial solution of $y'' + r(x)\,y = 0$ on $(0, \infty)$, where $r(x) > 0$.
If $\varphi(x) > 0$ on $(0, a)$ for some positive number a, and if there is a point
$x_0 \in (0, a)$ where $\varphi'(x_0) < 0$, prove that φ vanishes at some point $x_1 > x_0$.

2.12   Determine which equations have oscillatory solutions on $(0, \infty)$:

(a) $y'' + (\sin^2 x + 1)y = 0$.
(b) $y'' - x^2 y = 0$.
(c) $y'' + \dfrac{1}{x}y = 0$.

2.13   Find the general solution of Bessel's equation of order $1/2$, and determine the
zeros of each independent solution.

2.14   If f is continuous on $[0, \infty)$ and $\lim_{x \to \infty} f(x) = 0$, prove that for any
positive number ε the solutions of $y'' + (\varepsilon + f(x))y = 0$ are oscillatory.

2.15   Prove that any solution of Airy's equation $y'' + xy = 0$ has an infinite number
of zeros on $(0, \infty)$, and at most one zero on $(-\infty, 0)$.

2.3   Self-Adjoint Differential Operator

Going back to the general form of the linear second order differential Eq. (2.1), in
slightly modified notation,

$$p(x)y'' + q(x)y' + r(x)y = 0, \tag{2.21}$$

we now wish to investigate the orthogonality properties of its solutions. This means
we should look for the C^2 solutions of (2.21) which lie in the inner product space
$\mathcal{L}^2$. Equation (2.21) can be written in the form

$$Ly = 0,$$

where

$$L = p(x)\frac{d^2}{dx^2} + q(x)\frac{d}{dx} + r(x), \tag{2.22}$$

is a linear differential operator of second order from $C^2(I)$ to $C(I)$. By setting $I = [a, b]$, $C^2(I)$ becomes a subspace of $\mathcal{L}^2(I)$ on which the inner product $\langle \cdot, \cdot \rangle$ is well defined, and $Ly \in \mathcal{L}^2(I)$ whenever $y \in C^2(I)$.

To motivate the discussion, it would help at this point to recall some of the notions of linear algebra. A linear operator in a vector space X is a mapping

$$A : X \to X$$

which satisfies

$$A(a\mathbf{x} + b\mathbf{y}) = aA\mathbf{x} + bA\mathbf{y} \quad \text{for all } a, b \in \mathbb{F}, \mathbf{x}, \mathbf{y} \in X.$$

If an inner product is defined on X, the *adjoint* of A, if it exists, is the operator A' which satisfies

$$\langle A\mathbf{x}, \mathbf{y} \rangle = \langle \mathbf{x}, A'\mathbf{y} \rangle \quad \text{for all } \mathbf{x}, \mathbf{y} \in X.$$

If $A' = A$, then A is said to be *self-adjoint.*

If X is a finite dimensional inner product space, such as $\mathbb{C}^n$ over $\mathbb{C}$, we know that any linear operator is represented, with respect to the orthonormal basis $\{\mathbf{e}_i : 1 \leq i \leq n\}$, by the matrix

$$A = \begin{pmatrix} a_{11} & \cdots & a_{1n} \\ \vdots & & \vdots \\ a_{n1} & \cdots & a_{nn} \end{pmatrix} = (a_{ij}).$$

Its adjoint is given by

$$A' = \begin{pmatrix} \bar{a}_{11} & \cdots & \bar{a}_{n1} \\ \vdots & & \vdots \\ \bar{a}_{1n} & \cdots & \bar{a}_{nn} \end{pmatrix} = (\bar{a}_{ji}) = \bar{A}^T,$$

where $\bar{A}$ is the *complex conjugate of* A,

$$\bar{A} = \begin{pmatrix} \bar{a}_{11} & \cdots & \bar{a}_{1n} \\ \vdots & & \vdots \\ \bar{a}_{n1} & \cdots & \bar{a}_{nn} \end{pmatrix},$$

and A^T is its *transpose*,

$$A^T = \begin{pmatrix} a_{11} & \cdots & a_{n1} \\ \vdots & & \vdots \\ a_{1n} & \cdots & a_{nn} \end{pmatrix}.$$

A complex number a is called an *eigenvalue* of A if there is a non-zero vector $\mathbf{x} \in X$ such that

$$A\mathbf{x} = a\mathbf{x},$$

and $\mathbf{x}$ in this case is called an *eigenvector* of the operator A associated with the eigenvalue a. From linear algebra (see, for example, [11]) we know that, if A is a self-adjoint (or *Hermitian*) matrix, where $\overline{A}^T = A$, then

  (i) the eigenvalues of A are all real numbers,

 (ii) the eigenvectors of A corresponding to distinct eigenvalues are orthogonal,

(iii) the eigenvectors of A form a basis of X.

In order to extend these results to the differential operator $L : C^2(I) \cap \mathcal{L}^2(I) \to \mathcal{L}^2(I)$, our first task is to obtain the form of the *adjoint* of L as defined in Eq. (2.22). We assume, to start with, that the coefficients p, q and r are C^2 functions on I. Denoting the adjoint of L by L', we have, by definition of L',

$$\langle Lf, g \rangle = \langle f, L'g \rangle \quad \text{for all } f, g \in C^2(I) \cap \mathcal{L}^2(I). \tag{2.23}$$

With $I = [a, b]$, we use integration by parts to manipulate the left-hand side of (2.23) in order to shift the differential operator from f to g. Thus

$$\langle Lf, g \rangle = \int_a^b (pf'' + qf' + rf)\bar{g}\,dx$$

$$= pf'\bar{g}\big|_a^b - \int_a^b f'(p\bar{g})'dx + qf\bar{g}\big|_a^b - \int_a^b f(q\bar{g})'dx + \int_a^b fr\bar{g}\,dx$$

$$= [pf'\bar{g} - f(p\bar{g})']\big|_a^b + \int_a^b f(p\bar{g})''dx + qf\bar{g}\big|_a^b - \int_a^b f(q\bar{g})'dx$$

$$+ \int_a^b fr\bar{g}\,dx$$

$$= \langle f, (\bar{p}g)'' - (\bar{q}g)' + \bar{r}g \rangle + [p(f'\bar{g} - f\bar{g}') + (q - p')f\bar{g}]\big|_a^b .$$

We therefore have, for all $f, g \in \mathcal{L}^2(I) \cap C^2(I)$,

$$\langle Lf, g \rangle = \langle f, L^*g \rangle + [p(f'\bar{g} - f\bar{g}') + (q - p')f\bar{g}]\big|_a^b , \qquad (2.24)$$

where

$$L^*g = (\bar{p}g)'' - (\bar{q}g)' + \bar{r}g$$
$$= \bar{p}g'' + (2\bar{p}' - \bar{q})g' + (\bar{p}'' - \bar{q}' + \bar{r})g.$$

The differential operator

$$L^* = \bar{p}\frac{d^2}{dx^2} + (2\bar{p}' - \bar{q})\frac{d}{dx} + (\bar{p}'' - \bar{q}' + \bar{r})$$

is called the *formal adjoint* of L, and L is said to be *formally self-adjoint* when $L^* = L$, that is, when

$$\bar{p} = p, \ 2\bar{p}' - \bar{q} = q, \ \bar{p}'' - \bar{q}' + \bar{r} = r.$$

These three equations are satisfied if, and only if, the functions p, q, and r are real and $q = p'$. In that case

$$Lf = pf'' + p'f' + rf$$
$$= (pf')' + rf.$$

Thus, when L is formally self-adjoint, it has the form

$$L = \frac{d}{dx}\left(p\frac{d}{dx}\right) + r,$$

and Eq. (2.24) is reduced to

$$\langle Lf, g \rangle = \langle f, Lg \rangle + p(f'\bar{g} - f\bar{g}')\big|_a^b . \qquad (2.25)$$

Comparing Eq. (2.25) to Eq. (2.23) we now see that the formally self-adjoint operator L is self-adjoint if

$$p(f'\bar{g} - f\bar{g}')\big|_a^b = 0 \ \text{ for all } f, g \in C^2(I). \qquad (2.26)$$

It is worth noting at this point that, when $q = p'$, the term $\bar{p}'' - \bar{q}'$ in the expression for L^* drops out, hence the continuity of p'' and q' is no longer required.

The homogeneous differential equation

$$Lu + \lambda u = 0 \qquad\qquad (2.27)$$

will now be the subject of our study. When $u = 0$ this equation, of course, is satisfied for every value of λ. When $u \neq 0$, it may be satisfied for certain values of λ. These are the *eigenvalues* of $-L$. Any function $u \neq 0$ in $C^2(I)$ which satisfies Eq. (2.27) for some complex number λ is an *eigenfunction* of $-L$ corresponding to the eigenvalue λ. Occasionally we also refer to the eigenvalues and eigenfunctions of $-L$ as eigenvalues and eigenfunctions of Eq. (2.27). Since this equation is homogeneous, the eigenfunctions of $-L$ are determined up to a multiplicative constant. When appropriate boundary conditions are added to Eq. (2.27), the resulting system is called a *Sturm-Liouville eigenvalue problem*, which will be discussed in the next section. Clearly, $-L$ is self-adjoint if, and only if, L is self-adjoint. The reason we look for the eigenvalues of $-L$ rather than L is that, as it turns out, L has negative eigenvalues when p is positive (see Example 2.15 below). The following theorem summarizes the results we have obtained so far, as it generalizes properties (i) and (ii) above from finite to infinite dimensional space.

Theorem 2.14 *Let L be a linear differential operator of second order defined on $C^2(I) \cap \mathcal{L}^2(I)$ by*

$$Lu = p(x)u'' + q(x)u' + r(x)u, \quad x \in I,$$

where $I = [a, b]$, $p \in C^1(I)$, and $q, r \in C(I)$. Then

(i) *L is formally self-adjoint, that is, $L^* = L$, if the coefficients p, q, and r are real and $q = p'$.*

(ii) *L is self-adjoint, that is, $L' = L$, if L is formally self-adjoint and Eq. (2.26) is satisfied.*

(iii) *If L is self-adjoint, then the eigenvalues of the equation*

$$Lu + \lambda u = 0$$

are all real and any pair of eigenfunctions associated with distinct eigenvalues are orthogonal in $\mathcal{L}^2(I)$.

Proof We have already proved (i) and (ii). To prove (iii), suppose $\lambda \in \mathbb{C}$ is an eigenvalue of $-L$. Then there is a function $f \in C^2(I)$, $f \neq 0$, such that

$$Lf + \lambda f = 0$$

$$\Rightarrow \lambda \|f\|^2 = \langle \lambda f, f \rangle = - \langle Lf, f \rangle.$$

Since L is self-adjoint,

$$- \langle Lf, f \rangle = - \langle f, Lf \rangle = \langle f, \lambda f \rangle = \bar{\lambda} \, \| f \|^2 .$$

Hence $\bar{\lambda} \, \| f \|^2 = \lambda \, \| f \|^2$. Because $\| f \| \neq 0, \bar{\lambda} = \lambda$.

If μ is another eigenvalue of $-L$ associated with the eigenfunction $g \in \mathcal{L}^2(I) \cap C^2(I)$, then

$$\lambda \langle f, g \rangle = - \langle Lf, g \rangle = - \langle f, Lg \rangle = \mu \langle f, g \rangle$$

$$(\lambda - \mu) \langle f, g \rangle = 0$$

$$\lambda \neq \mu \implies \langle f, g \rangle = 0.$$

$\square$

Example 2.15 The differential operator $- \dfrac{d^2}{dx^2}$ is formally self-adjoint with $p = -1$ and $r = 0$. To determine its eigenfunctions in $C^2(0, \pi)$, we have to solve the equation

$$u'' + \lambda u = 0.$$

Let us first consider the case where $\lambda > 0$. The general solution of the equation is given by

$$u(x) = c_1 \cos \sqrt{\lambda} x + c_2 \sin \sqrt{\lambda} x. \tag{2.28}$$

Under the boundary conditions

$$u(0) = u(\pi) = 0$$

Equation (2.26) is satisfied, and $- \dfrac{d^2}{dx^2}$ is, in fact, self-adjoint. Applying the boundary conditions to (2.28), we obtain

$$u(0) = c_1 = 0$$

$$u(\pi) = c_2 \sin \sqrt{\lambda} \pi = 0 \implies \sqrt{\lambda} \pi = n\pi \implies \lambda = n^2, \ n \in \mathbb{N}.$$

Thus the eigenvalues of $- \dfrac{d^2}{dx^2}$ are given by the sequence

$$(n^2 : n \in \mathbb{N}) = (1, 4, 9, \cdots),$$

and the corresponding eigenfunctions are $(\sin x, \sin 2x, \sin 3x, \cdots)$.

Observe that we chose $c_2 = 1$ for the sake of simplicity, since the boundary conditions as well as the eigenvalue equation are all homogeneous. We could also divide each eigenfunction by its norm $\|\sin nx\| = \sqrt{\int_0^\pi \sin^2 x \, dx} = \sqrt{\pi/2}$ to obtain the normalized eigenfunctions

$$\left(\sqrt{2/\pi} \sin nx : n \in \mathbb{N} \right).$$

If $\lambda = 0$, the solution of the differential equation is given by $c_1 x + c_2$, and if $\lambda < 0$ it is given by $c_1 \cosh \sqrt{-\lambda} x + c_2 \sinh \sqrt{-\lambda} x$. In either case the application of the boundary conditions at $x = 0$ and $x = \pi$ leads to the conclusion that $c_1 = c_2 = 0$. Since the trivial solution is not admissible as an eigenfunction, we do not have any eigenvalues in the interval $(-\infty, 0]$. The eigenvalues $\lambda_n = n^2$ are real numbers, in accordance with Theorem 2.14, and the eigenfunctions $u_n(x) = \sin nx$ are orthogonal in $\mathcal{L}^2(0, \pi)$ since, for all $n \neq m$,

$$\int_0^\pi \sin nx \sin mx \, dx = \frac{1}{2} \left[\frac{\sin(n-m)x}{n-m} - \frac{\sin(n+m)x}{n+m} \right]\Big|_0^\pi = 0.$$

Sometimes it is not possible to determine the eigenvalues of a system exactly, as the next example shows.

Example 2.16 Consider the same equation

$$u'' + \lambda u = 0$$

on the interval $[0, l]$, under the separated boundary conditions

$$u(0) = 0, \quad hu(l) + u'(l) = 0,$$

where h is a positive constant.

It is straightforward to check that this system has no eigenvalues in $(-\infty, 0]$. When $\lambda > 0$ the general solution is

$$u(x) = c_1 \cos \sqrt{\lambda} x + c_2 \sin \sqrt{\lambda} x.$$

The first boundary condition implies $c_1 = 0$, and the second yields

$$c_2(h \sin \sqrt{\lambda} l + \sqrt{\lambda} \cos \sqrt{\lambda} l) = 0.$$

Since c_2 cannot be 0, otherwise we do not get an eigenfunction, we must have

$$h \sin \sqrt{\lambda} l + \sqrt{\lambda} \cos \sqrt{\lambda} l = 0.$$

Therefore λ has to satisfy

$$\tan \sqrt{\lambda}\, l = -\frac{\sqrt{\lambda}}{h}.$$

Setting $\alpha = \sqrt{\lambda}\, l$, we see that α is determined by the equation

$$\tan \alpha = -\frac{\alpha}{hl}.$$

This is a transcendental equation which is solved graphically by the points of intersection of the graphs of

$$y = \tan \alpha \quad \text{and} \quad y = -\frac{\alpha}{hl},$$

as shown by the sequence (α_n) in Fig. 2.2, where $(n - 1/2)\,\pi \; < \alpha_n \; < n\pi$ for all $n \in \mathbb{N}$ and α_n approaches $(n - 1/2)\,\pi$ asyptotically as $n \to \infty$.

The eigenvalues and eigenfunctions of the problem are therefore given by

$$\lambda_n = \left(\frac{\alpha_n}{l}\right)^2,$$

$$u_n(x) = \sin\left(\frac{\alpha_n}{l}x\right), \; 0 \le x \le l, \; n \in \mathbb{N}.$$

Fig. 2.2 The sequence α_n

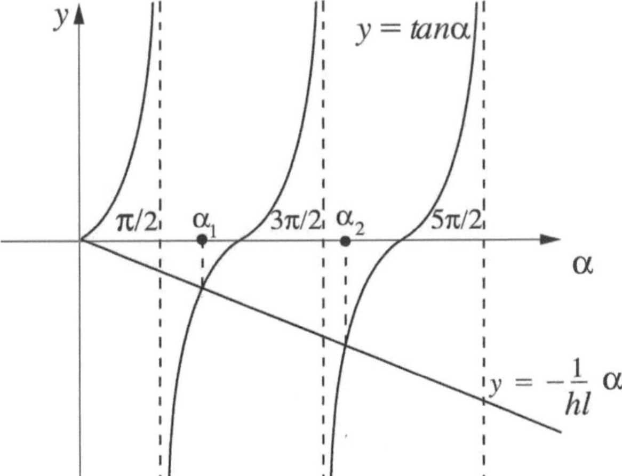

The following example shows that, if $p' \neq q$ on I, the operator L in Theorem 2.14 may be transformed to a formally self-adjoint operator when multiplied by a suitable function.

Example 2.17 Let

$$L = p(x)\frac{d^2}{dx^2} + q(x)\frac{d}{dx} + r(x), \ x \in I = [a, b].$$

If the coefficients p, q, and r are real, p does not vanish on I, but $q \neq p'$ then L is not formally self-adjoint. Assuming, without loss of generality, that $p > 0$ on $[a, b]$, we can multiply L by a function ρ, to be determined, such that the operator

$$\tilde{L} = \rho L = \rho p\frac{d^2}{dx^2} + \rho q\frac{d}{dx} + \rho r,$$

is formally self-adjoint. That implies ρ satisfies

$$\rho q = (\rho p)' = \rho' p + \rho p',$$

which is a first order differential equation in ρ, whose solution is

$$\rho(x) = \frac{c}{p(x)} \exp\left(\int_a^x \frac{q(t)}{p(t)}dt\right).$$

The integration constant c is set equal to 1 so that $\rho = 1$ when $q = p'$, i.e. when L is formally self-adjoint. Therefore

$$\rho(x) = \frac{1}{p(x)} \exp\left(\int_a^x \frac{q(t)}{p(t)}dt\right), \tag{2.29}$$

which is a positive C^2 function on $[a, b]$.

the result of Example 2.17 allows us to generalize part (iii) of Theorem 2.14. If the differential operator $L = p\left(d^2/dx^2\right) + q\left(d/dx\right) + r$ is not formally self-adjoint because $q \neq p'$, and $p > 0$, the eigenvalue equation

$$Lu + \lambda u = 0$$

can be multiplied by the positive C^1 function ρ defined by (2.29) to give the equivalent equation

$$\rho Lu + \lambda \rho u = 0, \tag{2.30}$$

where ρL is now formally self-adjoint. With $\rho > 0$, Eq. (2.26) is equivalent to

$$\rho p(f'\bar{g} - f\bar{g}')\big|_a^b = 0.$$

This makes the operator ρL self-adjoint. If $u \in \mathcal{L}_\rho^2$ is an eigenfunction of L associated with the eigenvalue λ, we can write

$$\begin{aligned}
\lambda \|u\|_\rho^2 &= \langle \lambda \rho u, u \rangle \\
&= \langle -\rho L u, u \rangle \\
&= \langle u, -\rho L u \rangle \\
&= \langle u, \lambda u \rangle_\rho \\
&= \bar{\lambda} \|u\|_\rho^2,
\end{aligned}$$

which implies that λ is a real number. Furthermore, if $v \in \mathcal{L}_\rho^2$ is another eigenfunction of L associated with the eigenvalue μ, then

$$\begin{aligned}
(\lambda - \mu) \langle u, v \rangle_\rho &= \lambda \langle \rho u, v \rangle - \mu \langle \rho u, v \rangle \\
&= \langle \lambda \rho u, v \rangle - \langle u, \mu \rho v \rangle \\
&= \langle -\rho L u, v \rangle - \langle u, -\rho L v \rangle = 0
\end{aligned}$$

because ρL is self-adjoint. Thus, if $\lambda \neq \mu$, then u is orthogonal to v in $\mathcal{L}_\rho^2$.

We have therefore proved the following extension of Theorem 2.14(iii):

Corollary 2.18 *If L is a self-adjoint linear operator and ρ is a positive and continuous function on the interval I, then the eigenvalues of the equation*

$$Lu + \lambda \rho u = 0$$

are all real and any pair of eigenfunctions associated with distinct eigenvalues are orthogonal in $\mathcal{L}_\rho^2(I)$.

Remark 2.19

1. In this corollary the eigenvalues and eigenfunctions of the equation $Lu + \lambda \rho u = 0$ are, in fact, the eigenvalues and eigenfunctions of the operator $-\rho^{-1} L$.

2. The weight function ρ being continuous and positive on the interval $I = [a, b]$, its minimum value α and its maximum value β satisfy

$$0 < \alpha \leq \rho(x) \leq \beta < \infty.$$

This implies

$$\sqrt{\alpha} \|u\| \leq \|u\|_\rho \leq \sqrt{\beta} \|u\|,$$

and therefore $\|u\|_\rho < \infty$ if, and only if, $\|u\| < \infty$. The two norms are said to be *equivalent*, and the two spaces $\mathcal{L}^2_\rho(a,b)$ and $\mathcal{L}^2(a,b)$ clearly contain the same functions, though they have different inner products.

3. Nothing in the proof of this corollary requires L to be a linear second order differential operator. In fact, the result is true for any self-adjoint linear operator on an inner product space.

Example 2.20 Find the eigenfunctions and eigenvalues of the boundary-value problem

$$x^2 y'' + xy' + \lambda y = 0, \ 1 < x < b, \tag{2.31}$$

$$y(1) = y(b) = 0. \tag{2.32}$$

Solution Equation (2.31) is a Cauchy-Euler equation whose solutions have the form x^m. Substituting into the differential equation leads to

$$m(m-1) + m + \lambda = 0$$

$$\Rightarrow m = \pm i\sqrt{\lambda}.$$

Assuming $\lambda > 0$, we have

$$x^{i\sqrt{\lambda}} = e^{i\sqrt{\lambda}\log x} = \cos(\sqrt{\lambda}\log x) + i\sin(\sqrt{\lambda}\log x),$$

and the general solution of Eq. (2.31) is given by

$$y(x) = c_1 \cos(\sqrt{\lambda}\log x) + c_2 \sin(\sqrt{\lambda}\log x).$$

Applying the boundary conditions (2.32),

$$y(1) = c_1 = 0, \ \ y(b) = c_2 \sin(\sqrt{\lambda}\log b) = 0.$$

Since we cannot have both constants c_1 and c_2 vanish, this implies

$$\sin(\sqrt{\lambda}\log b) = 0$$

$$\Rightarrow \sqrt{\lambda}\log b = n\pi, \ n \in \mathbb{N}.$$

Thus the eigenvalues of the boundary value problem are given by the sequence of positive real numbers

$$\lambda_n = \left(\frac{n\pi}{\log b}\right)^2, \ \ n \in \mathbb{N},$$

and the corresponding sequence of eigenfunctions is

$$y_n(x) = \sin\left(\frac{n\pi}{\log b}\log x\right).$$

Observe here that the differential operator

$$x^2\frac{d^2}{dx^2} + x\frac{d}{dx}$$

is not formally self-adjoint, but it becomes so after multiplication by the weight function

$$\rho(x) = \frac{1}{x^2}\exp\left(\int_1^x \frac{1}{t}dt\right) = \frac{1}{x}.$$

The resulting self-adjoint operator is then

$$L = x\frac{d^2}{dx^2} + \frac{d}{dx} = \frac{d}{dx}\left(x\frac{d}{dx}\right).$$

The eigenfunctions $(y_n : n \in \mathbb{N})$ of this problem are in fact eigenfunctions of $-\rho^{-1}L = -xL$, which are orthogonal in $\mathcal{L}_\rho^2(1, b)$, not $\mathcal{L}^2(1, b)$. Indeed,

$$\langle y_m, y_n \rangle_\rho = \int_1^b \sin\left(\frac{m\pi}{\log b}\log x\right)\sin\left(\frac{n\pi}{\log b}\log x\right)\frac{1}{x}dx$$

$$= \frac{\log b}{\pi}\int_0^\pi \sin m\xi \sin n\xi\, d\xi$$

$$= 0 \ \text{ for all } m \neq n.$$

We leave it as an exercise to show that this problem has no eigenvalues in the interval $(-\infty, 0]$.

Exercises

2.16 Given $L = \dfrac{d}{dx}\left(p\dfrac{d}{dx}\right) + r$, prove the *Lagrange identity*

$$uLv - vLu = [p(uv' - vu')]'.$$

The integral of this identity,

$$\int_a^b (uLv - vLu)\,dx = \left[p(uv' - vu')\right]\Big|_a^b,$$

is known as *Green's formula*.

2.17 Determine the eigenfunctions and eigenvalues of the differential operators

(a) $\dfrac{d^2}{dx^2} : C^2(0, \infty) \to C(0, \infty),$

(b) $\dfrac{d^2}{dx^2} : \mathcal{L}^2(0, \infty) \cap C^2([0, \infty)) \to \mathcal{L}^2(0, \infty).$

2.18 Prove that the differential operator $-\dfrac{d^2}{dx^2} : \mathcal{L}^2(0, \pi) \cap C^2([0, \pi]) \to \mathcal{L}^2(0, \pi)$ is self-adjoint under the boundary conditions $u(0) = u'(\pi) = 0$. Verify that its eigenvalues and eigenfunctions satisfy Theorem 2.14(iii).

2.19 Put each of the following differential operators in the form $p\dfrac{d^2}{dx^2} + p'\dfrac{d}{dx} + r$, with $p > 0$, by multiplying by a suitable weight function:

(a) $x^2\dfrac{d^2}{dx^2}, \quad x > 0,$

(b) $\dfrac{d^2}{dx^2} - x\dfrac{d}{dx}, \quad x \in \mathbb{R},$

(c) $\dfrac{d^2}{dx^2} - x^2\dfrac{d}{dx}, \quad x > 0,$

(d) $x^2\dfrac{d^2}{dx^2} + x\dfrac{d}{dx} + (x^2 - \lambda), \quad x > 0.$

2.20 Determine the positive eigenvalues and the eigenfunctions of $u'' + \lambda u = 0$ on $(0, l)$ subject to the boundary conditions $u(0) = 0$, $hu(l) + u'(l) = 0$, where $h \le 0$. With reference to Fig. 2.2 in Example 2.16, discuss the asymptotic distribution of α_n on $[0, \infty)$ in this case.

2.21 Put the eigenvalue equation $u'' + 2u' + \lambda u = 0$ in the standard form $Lu + \lambda \rho u = 0$, where L is self-adjoint, then find its eigenvalues and eigenfunctions on $[0, 1]$ subject to the boundary conditions $u(0) = u(1) = 0$. Verify that the result agrees with Corollary 2.19.

2.22 Determine the eigenfunctions and eigenvalues of the equation $x^2 u'' + \lambda u = 0$ on $[1, e]$, subject to the boundary conditions $u(1) = u(e) = 0$.

2.4 The Sturm-Liouville Problem

In Sect. 2.3 we saw how a linear differential equation may be posed as an eigenvalue problem for a differential operator in an inner product space of functions. The notion of self-adjointness was defined by drawing on the analogy with linear transformations in finite dimensional space, and we saw how the first two properties of a self-adjoint matrix, namely that its eigenvalues are real and its eigenvectors orthogonal, translated into Theorem 2.14. The third property, that its eigenvectors span the space $\mathbb{R}^n$ or $\mathbb{C}^n$, is in fact an assertion that the matrix has orthogonal eigenvectors as well as a statement on their number. This section is concerned with deriving the corresponding result for differential operators, which is clearly expressed by Theorems 2.27, the prime result of the chapter.

Let L be the formally self-adjoint differential operator

$$L = \frac{d}{dx}\left(p(x)\frac{d}{dx}\right) + r(x). \tag{2.33}$$

The eigenvalue equation

$$Lu + \lambda\rho(x)u = 0, \ x \in (a,b), \tag{2.34}$$

where p', r, and ρ are all real continuous functions on (a,b), subject to the separated homogeneous boundary conditions

$$\alpha_1 u(a) + \alpha_2 u'(a) = 0, \ |\alpha_1| + |\alpha_2| > 0, \tag{2.35}$$

$$\beta_1 u(b) + \beta_2 u'(b) = 0, \ |\beta_1| + |\beta_2| > 0, \tag{2.36}$$

where α_i and β_i are real constants, is called a *Sturm-Liouville eigenvalue problem*, or *SL problem* for short. This is a homogeneous system of equations which is always solved by $u \equiv 0$, but it may have other (nontrivial) solutions for certain values of λ, as we have seen in Examples 2.15 and 2.20. These are the eigenvalues of the problem, and the corresponding solutions are the eigenfunctions.

When the interval (a,b) is bounded, $p', r \in C([a,b])$, and $p(x) \neq 0$ for any $x \in [a,b]$, the system of Eqs. (2.34)–(2.36) is called a *regular* SL problem, otherwise it is *singular.* In the regular problem there is no loss of generality in assuming that $p(x) > 0$ on $[a,b]$, and we know from Corollary 2.18 that, if they exist, the eigenvalues of Eq. (2.34) are real and the eigenfunctions are orthogonal in $\mathcal{L}^2_\rho(a,b)$, where the weight function ρ is positive and continuous on $[a,b]$. The solutions of the SL problem are clearly the eigenfunctions of the operator $-\rho^{-1}L$ in $C^2([a,b])$ which satisfy the boundary conditions (2.35) and (2.36).

Now we show that the regular SL problem not only has solutions, but that there are enough of them to span $\mathcal{L}^2_\rho(a,b)$. This fundamental result will be proved in stages. Assuming, for the sake of simplicity, that $\rho(x) = 1$, we first construct *Green's function* for the operator L under the boundary conditions (2.35) and (2.36). This choice of ρ avoids some distracting details in order to focus on the general principle. We then use Green's function to arrive at an integral expression for L^{-1}, the inverse of L. The eigenfunctions of $-L$,

$$Lu + \lambda u = 0,$$

coincide with those of $-L^{-1}$,

$$L^{-1}u + \frac{1}{\lambda}u = 0,$$

and the corresponding eigenvalues are $1/\lambda$. We arrive at the desired conclusions by analysing the spectral properties of L^{-1}, that is, its eigenvalues and eigenfunctions. We shall have more to say about the singular problem at the end of this section.

2.4.1   Existence of Eigenfunctions

Green's function for the self-adjoint differential operator

$$L = p\frac{d^2}{dx^2} + p'\frac{d}{dx} + r,$$

under the boundary conditions

$$\alpha_1 u(a) + \alpha_2 u'(a) = 0, \ |\alpha_1| + |\alpha_2| > 0,$$
$$\beta_1 u(b) + \beta_2 u'(b) = 0, \ |\beta_1| + |\beta_2| > 0,$$

is a function

$$G : [a, b] \times [a, b] \to \mathbb{R}$$

with the following properties:

1. G is symmetric, in the sense that

$$G(x, \xi) = G(\xi, x) \ \text{ for all } x, \xi \in [a, b],$$

   and G satisfies the above boundary conditions in each variable x and ξ on (a, b).
2. G is a continuous function on the square $[a, b] \times [a, b]$ and of class C^2 on $[a, b] \times [a, b] \backslash \{(x, \xi) : x = \xi\}$, that is, excluding the diagonal $x = \xi$ of the square, where it satisfies the differential equation

$$L_x G(x, \xi) = p(x)G_{xx}(x, \xi) + p'(x)G_x(x, \xi) + r(x)G(x, \xi) = 0.$$

3. The derivative $G_x = \partial G/\partial x$ has a jump discontinuity at $x = \xi$ given by (Fig. 2.3)

$$\frac{\partial G}{\partial x}(x^+, x) - \frac{\partial G}{\partial x}(x^-, x) = \lim_{\delta \to 0^+} \left[\frac{\partial G}{\partial x}(x + \delta, x) - \frac{\partial G}{\partial x}(x - \delta, x) \right]$$

Fig. 2.3 Green's function

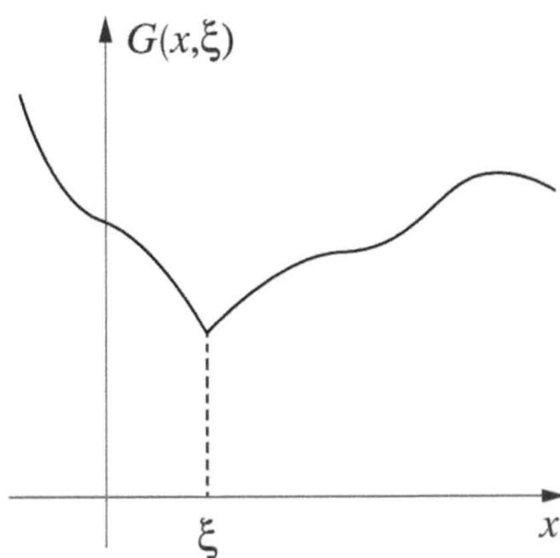

$$= \frac{1}{p(x)}, \qquad x \in (a, b). \tag{2.37}$$

We assume that the homogeneous equation $Lu = 0$, under the separated boundary conditions (2.35) and (2.36), has only the trivial solution $u = 0$. This amounts to the assumption that 0 is not an eigenvalue of $-L$. There is no loss of generality in this assumption, for if $\tilde{\lambda}$ is a real number which is *not* an eigenvalue of $-L$ then we can define the operator

$$\tilde{L} = L + \tilde{\lambda} = p\frac{d^2}{dx^2} + p'\frac{d}{dx} + \tilde{r},$$

where $\tilde{r}(x) = r(x) + \tilde{\lambda}$. Since $\tilde{\lambda}$ is not an eigenvalue of $-L$, 0 cannot be an eigenvalue of $\tilde{L}$. We can therefore assume, without loss of generality, that 0 is not an eigenvalue of L. That there are real numbers which are not eigenvalues of $-L$ follows from the following lemma.

Lemma 2.21 *The eigenvalues of $-L$ are bounded below by a real constant.*

Proof For any $u \in C^2([a, b])$,

$$\langle -Lu, u \rangle = \int_a^b [-(pu')'\,]\overline{u} - r\,|u|^2]dx$$

$$= \int_a^b [p\,|u'|^2 - r\,|u|^2]dx + p(a)\,u'(a)\,\overline{u}(a) - p(b)\,u'(b)\,\overline{u}(b).$$

If $u(a) = u(b) = 0$, then

$$\lambda \|u\|^2 = \int_a^b p(x) \left|u'(x)\right|^2 dx - \int_a^b r(x) |u(x)|^2] dx$$

$$\geq - \|u\|^2 \max \{|r(x)| : a \leq x \leq b\}, \tag{2.38}$$

and $\ell = -\max\{|r(x)| : a \leq x \leq b\}$ is a lower bound of λ.

If, on the other hand, u satisfies the separated boundary conditions (2.35) and (2.36), then the following dimensional argument shows that there can be no more than two linearly independent eigenfunctions of $-L$ with eigenvalues less than ℓ.

Seeking a contradiction, suppose λ_1, λ_2, and λ_3 are distinct eigenvalues of $-L$, each less than ℓ, and that u_1, u_2, and u_3 are the corresponding orthonormal eigenfunctions which satisfy

$$\alpha_1 u_i(a) + \alpha_2 u_i'(a) = 0,$$

$$\beta_1 u_i(b) + \beta_2 u_i'(b) = 0, \quad i = 1, 2, 3.$$

Each of the 6 vectors $\left(u_i(a), u_i'(a)\right)$ and $\left(u_i(b), u_i'(b)\right)$ lies in a one-dimensional subspace of $\mathbb{C}^2$, and so the three vectors

$$\mathbf{u}_i = \left(u_i(a), u_i'(a), u_i(b), u_i'(b)\right) \quad i = 1, 2, 3,$$

lie in a two-dimensional subspace of $\mathbb{C}^4$. Consequently there is a nontrivial linear combination of these vectors such that

$$c_1 \mathbf{u}_1 + c_2 \mathbf{u}_2 + c_3 \mathbf{u}_3 = \mathbf{0}.$$

But the function $v(x) = c_1 u_1(x) + c_2 u_2(x) + c_3 u_3(x)$ satisfies $v(a) = v(b) = 0$, so we have

$$\langle -Lv, v \rangle = \lambda_1 |c_1|^2 + \lambda_2 |c_2|^2 + \lambda_3 |c_3|^2 < \ell \|v\|,$$

in direct contradiction to the inequality (2.38). $\qquad\square$

Now we construct Green's function for the operator L under boundary conditions (2.35) and (2.36),

$$\alpha_1 u(a) + \alpha_2 u'(a) = 0, \ |\alpha_1| + |\alpha_2| > 0,$$

$$\beta_1 u(b) + \beta_2 u'(b) = 0, \ |\beta_1| + |\beta_2| > 0.$$

According to a standard existence theorem for second order differential equations [6], $Lu = 0$ has two unique (non-trivial) solutions v_1 and v_2 such that

$$v_1(a) = \alpha_2, \quad v_1'(a) = -\alpha_1,$$
$$v_2(b) = \beta_2, \quad v_2'(b) = -\beta_1.$$

Therefore v_1 satisfies the first boundary condition at a, and v_2 satisfies the second,

$$\alpha_1 v_1(a) + \alpha_2 v_1'(a) = 0,$$
$$\beta_1 v_2(b) + \beta_2 v_2'(b) = 0.$$

Clearly v_1 and v_2 are linearly independent, otherwise one would be a (non-zero) constant multiple of the other, and both would then satisfy $Lu = 0$ and the boundary conditions (2.35) and (2.36). But this would contradict the assumption that 0 is not an eigenvalue of L.

We define

$$G(x, \xi) = \begin{cases} c^{-1} v_1(\xi) v_2(x), & a \le \xi \le x \le b \\ c^{-1} v_1(x) v_2(\xi), & a \le x \le \xi \le b, \end{cases} \tag{2.39}$$

where

$$c = p(x)[v_1(x) v_2'(x) - v_1'(x) v_2(x)] = p(x) W(v_1, v_2)(x) \tag{2.40}$$

is a non-zero constant. This follows from the fact that neither p nor W vanishes on $[a, b]$, and the Lagrange identity (see Exercise 2.16)

$$[p(v_1 v_2' - v_1' v_2)]' = v_1 L v_2 - v_2 L v_1 = 0 \quad \text{on } [a, b].$$

It is straightforward to show that all the properties of $G(\xi, x)$ listed above are satisfied. Here we shall verify Eq. (2.37). Differentiating (2.39) and using the continuity of v_1' and v_2', we obtain

$$\frac{\partial G}{\partial x}(x^+, x) - \frac{\partial G}{\partial x}(x^-, x) = \frac{1}{c}[v_1(x) v_2'(x) - v_1'(x) v_2(x)]$$
$$= \frac{1}{p(x)}.$$

We now define the integral operator T on $C([a, b])$ by

$$(Tf)(x) = \int_a^b G(x, \xi) f(\xi) d\xi, \tag{2.41}$$

and show that the function Tf is of class $C^2([a, b])$ and solves the differential equation $Lu = f$. Rewriting (2.41) and differentiating,

$$(Tf)(x) = \int_a^x G(x, \xi) f(\xi) d\xi + \int_x^b G(x, \xi) f(\xi) d\xi,$$

$$(Tf)'(x) = \int_a^x G_x(x, \xi) f(\xi) d\xi + \int_x^b G_x(x, \xi) f(\xi) d\xi,$$

$$(Tf)''(x) = \int_a^x G_{xx}(x, \xi) f(\xi) d\xi + G_x(x, x^+) f(x)$$

$$+ \int_x^b G_{xx}(x, \xi) f(\xi) d\xi - G_x(x, x^-) f(x),$$

where we used the continuity of G and f at $\xi = x$ to obtain $(Tf)'(x)$. Using Eq. (2.37),

$$(Tf)''(x) = \int_a^x G_{xx}(x, \xi) f(\xi) d\xi + \int_x^b G_{xx}(x, \xi) f(\xi) d\xi + \frac{f(x)}{p(x)},$$

from which we conclude that $Tf \in C^2([a, b])$ and satisfies

$$L(Tf)(x) = p(x)(Tf)''(x) + p'(x)(Tf)'(x) + r(x)(Tf)(x)$$

$$= \int_a^x L_x G(x, \xi) f(\xi) d\xi + \int_x^b L_x G(x, \xi) f(\xi) d\xi + f(x)$$

$$= f(x),$$

in view of the fact that $L_x G(x, \xi) = 0$ for all $\xi \neq x$.

From property (1) of G it is clear that Tf satisfies the boundary conditions (2.35) and (2.36). On the other hand, if $u \in C^2([a, b])$ satisfies the boundary conditions (2.35) and (2.36), then we can integrate by parts and use the continuity of p, u, u' and the properties of G to obtain

$$T(Lu)(x) = \int_a^x G(x, \xi) Lu(\xi) d\xi + \int_x^b G(x, \xi) Lu(\xi) d\xi$$

$$= p(\xi) u(\xi) G_\xi(x, \xi) \big|_{x^-}^{x^+}$$

$$= u(x),$$

where we used Eq. (2.37) and the fact that u and G satisfy the separated boundary conditions. The integral operator T therefore serves to reduce the SL system of equations

$$Lu + \lambda u = 0,$$

$$\alpha_1 u(a) + \alpha_2 u'(a) = 0,$$

$$\beta_1 u(b) + \beta_2 u'(b) = 0,$$

to the single equivalent eigenvalue equation

$$Tu = \mu u,$$

where $\mu = -1/\lambda$. In other words, u is an eigenfunction of the SL problem associated with the eigenvalue λ if, and only if, it is an eigenfunction of T associated with the eigenvalue $-1/\lambda$. As would be expected, for a regular SL problem, the integral operator T is self-adjoint on $C \cap \mathcal{L}^2$ [6]. Our next task is to deduce the spectral properties of the SL operator L from those of the integral operator T.

The following example illustrates the method we have described for constructing Green's function in the special case where the operator L is $d^2/dx^2 + \omega$ on the interval $[0, 1]$.

Example 2.22 Let

$$u'' + \omega u = 0, \quad 0 < x < 1,$$

$$u(0) = u(1) = 0,$$

where $\omega > 0$. The general solution of the differential equation is

$$v(x) = c_1 \cos \sqrt{\omega} x + c_2 \sin \sqrt{\omega} x.$$

Noting that $\alpha_1 = \beta_1 = 1$ and $\alpha_2 = \beta_2 = 0$, we seek the solution v_1 which satisfies the following initial conditions at $x = 0$:

$$v_1(0) = \alpha_2 = 0, \quad v_1'(0) = -\alpha_1 = -1.$$

These conditions imply

$$c_1 = 0, \quad \sqrt{\omega} c_2 = -1.$$

Hence

$$v_1(x) = -\frac{1}{\sqrt{\omega}} \sin \sqrt{\omega} x.$$

Similarly, the solution v_2 which satisfies the conditions

$$v_2(1) = \beta_2 = 0, \ v_2'(1) = -\beta_1 = -1,$$

is readily found to be

$$v_2(x) = \frac{\sin \sqrt{\omega}}{\sqrt{\omega}} \cos \sqrt{\omega}x - \frac{\cos \sqrt{\omega}}{\sqrt{\omega}} \sin \sqrt{\omega}x$$

$$= \frac{-1}{\sqrt{\omega}} \sin \sqrt{\omega}\,(x - 1).$$

Green's function, according to (2.39), is given by

$$G(x, \xi) = \begin{cases} c^{-1} v_1(\xi)v_2(x), \ 0 \le \xi \le x \le 1, \\ c^{-1} v_1(x)v_2(\xi), \ 0 \le x \le \xi \le 1. \end{cases}$$

The constant c can be evaluated using Eq. (2.40), and is given by $\sin \sqrt{\omega}/\sqrt{\omega}$. Thus

$$G(\xi, x) = \begin{cases} \dfrac{1}{\sqrt{\omega} \sin \sqrt{\omega}} \sin \sqrt{\omega}\xi \cdot \sin \sqrt{\omega}(x - 1) \ 0 \le \xi \le x \le 1, \\ \dfrac{1}{\sqrt{\omega} \sin \sqrt{\omega}} \sin \sqrt{\omega}x \cdot \sin \sqrt{\omega}(\xi - 1) \ 0 \le x \le \xi \le 1. \end{cases}$$

and

$$G_x(x^+, x^-) - G_x(x^-, x) = \frac{1}{\sin \sqrt{k}}[\sin \sqrt{\omega}x^+ \cdot \cos \sqrt{\omega}(x^+ - 1)$$

$$- \cos \sqrt{\omega}x^- \cdot \sin \sqrt{\omega}\,(x^- - 1)] = 1.$$

The other properties of G can easily be checked. Note that $\{(n\pi)^2 : n \in \mathbb{N}\}$ are the exceptional values where the homogeneous equation $Lu = 0$ has a non-trivial solution under the given boundary conditions, i.e. where 0 is an eigenvalue for L, so $\omega \ne (n\pi)^2$ in the above expression for $G(x, \xi)$.

A set F of continuous functions on $[a, b]$ is said to be *equicontinuous* on $[a, b]$ if, given any $\varepsilon > 0$, there is a $\delta > 0$ which depends only on ε, such that

$$x, \xi \in [a, b], \ |x - \xi| < \delta \ \Rightarrow \ |f(x) - f(\xi)| < \varepsilon \ \text{ for any } f \in F.$$

This condition clearly implies that every member of an equicontinuous set of functions is a uniformly continuous function on $[a, b]$; but, more than that, the same δ works for *all* the functions in F. The set F is *uniformly bounded* if there is a positive number M such that

$$|f(x)| \le M \ \text{ for all } f \in F, x \in [a, b].$$

According to the *Ascoli-Arzela theorem* [15], if F is a uniformly bounded and equicontinuous set of functions from the compact interval $[a, b]$ into $\mathbb{R}$, then any sequence of functions $(f_n : n \in \mathbb{N})$ in F contains a subsequence which is uniformly convergent on $[a, b]$. By Theorem 1.14, the limit of f_n is necessarily a continuous function on $[a, b]$.

Lemma 2.23 *Let T be the integral operator defined by Eq. (2.41). The set of functions $\{Tu\}$, where $u \in C([a, b])$ and $\|u\| \leq 1$, is uniformly bounded and equicontinuous.*

Proof Since Green's function G is continuous on the closed square $[a, b] \times [a, b]$, $|G(x, \xi)|$ is uniformly continuous and bounded by some positive constant M. From (2.41) and the CBS inequality,

$$|Tu(x)| = |\langle G(x, \xi), u(\xi)\rangle| \leq M\sqrt{b - a}\,\|u\|,$$

hence the set $\{|Tu| : \|u\| \leq 1\}$ is bounded above by $M\sqrt{b - a}$. Since G is uniformly continuous on the square $[a, b] \times [a, b]$, given any $\varepsilon > 0$, there is a $\delta > 0$ such that

$$x_1, x_2 \in [a, b], \ |x_2 - x_1| < \delta \ \Rightarrow \ |G(x_2, \xi) - G(x_1, \xi)| < \varepsilon \ \text{ for all } \xi \in [a, b].$$

If u is continuous on $[a, b]$, then

$$|x_2 - x_1| < \delta \ \Rightarrow \ |Tu(x_2) - Tu(x_1)| \leq \varepsilon\sqrt{b - a}\,\|u\|.$$

Thus the functions $\{Tu : u \in C[a, b], \|u\| \leq 1\}$ are equicontinuous.   □

The *norm* of the operator T, denoted by $\|T\|$, is defined by

$$\|T\| = \sup\{\|Tu\| : u \in C([a, b]), \|u\| = 1\}. \tag{2.42}$$

From this it follows that, if $u \neq 0$, then

$$\|Tu\| = \|T(u/\|u\|)\|\,\|u\| \leq \|T\|\,\|u\|.$$

But since this inequality also holds when $u = 0$, it holds for all $u \in C([a, b])$. It can also be shown (see Exercise 2.33) that

$$\|T\| = \sup_{\|u\|=1} |\langle Tu, u\rangle|. \tag{2.43}$$

Now we can state our first existence theorem for the eigenvalue problem under consideration.

Theorem 2.24 *Either $\|T\|$ or $-\|T\|$ is an eigenvalue of T.*

Proof $\|T\| = \sup\{|\langle Tu, u\rangle| : u \in C([a, b]), \|u\| = 1\}$ and $\langle Tu, u\rangle$ is a real number, since $\langle Tu, u\rangle = \langle u, Tu\rangle$, hence either $\|T\| = \sup \langle Tu, u\rangle$ or $\|T\| =$

$-\inf \langle Tu, u \rangle$, $\|u\| = 1$. Suppose $\|T\| = \sup \langle Tu, u \rangle$. Then there is a sequence of functions $u_k \in C([a, b])$, with $\|u_k\| = 1$, such that

$$\langle Tu_k, u_k \rangle \to \|T\| \quad \text{as } k \to \infty.$$

The sequence (Tu_k) is uniformly bounded and equicontinuous, so we can use the Ascoli-Arzela theorem to conclude that it has a subsequence (Tu_{k_j}) which is uniformly convergent on $[a, b]$ to a continuous function φ_0. We now prove that φ_0 is an eigenfunction of T corresponding to the eigenvalue $\mu_0 = \|T\|$.

As $j \to \infty$ we have

$$\sup_{x \in [a,b]} \left| Tu_{k_j}(x) - \varphi_0(x) \right| \to 0,$$

which implies

$$\left\| Tu_{k_j} - \varphi_0 \right\| \to 0, \tag{2.44}$$

and therefore $\left\| Tu_{k_j} \right\| \to \|\varphi_0\|$. Moreover, because $\langle Tu_{k_j}, u_{k_j} \rangle \to \mu_0$,

$$\left\| Tu_{k_j} - \mu_0 u_{k_j} \right\|^2 = \left\| Tu_{k_j} \right\|^2 + \mu_0^2 - 2\mu_0 \langle Tu_{k_j}, u_{k_j} \rangle \to \|\varphi_0\|^2 - \mu_0^2, \tag{2.45}$$

hence $\|\varphi_0\|^2 \geq \mu_0^2 > 0$, and the function φ_0 cannot vanish identically on $[a, b]$. From the inequality $\left\| Tu_{k_j} \right\|^2 \leq \|T\|^2 \left\| u_{k_j} \right\|^2 = \mu_0^2$ and (2.45) it also follows that

$$0 \leq \left\| Tu_{k_j} - \mu_0 u_{k_j} \right\|^2 \leq 2\mu_0^2 - 2\mu_0 \langle Tu_{k_j}, u_{k_j} \rangle.$$

But $\langle Tu_{k_j}, u_{k_j} \rangle \to \mu_0$, therefore

$$\left\| Tu_{k_j} - \mu_0 u_{k_j} \right\| \to 0. \tag{2.46}$$

Now we can write

$$0 \leq \left\| T\varphi_0 - \mu_0\varphi_0 \right\|$$
$$\leq \left\| T\varphi_0 - T(Tu_{k_j}) \right\| + \left\| T(Tu_{k_j}) - \mu_0 Tu_{k_j} \right\| + \left\| \mu_0 Tu_{k_j} - \mu_0\varphi_0 \right\|.$$

In the limit as $j \to \infty$, using the inequality $\|Tu\| \leq \|T\| \|u\|$ together with (2.44) and (2.46), we conclude that $\left\| T\varphi_0 - \mu_0\varphi_0 \right\| = 0$. Since $T\varphi_0 - \mu_0\varphi_0$ is continuous, this implies $T\varphi_0(x) = \mu_0\varphi_0(x)$ for all $x \in [a, b]$.

If $\|T\| = -\inf \langle Tu, u \rangle$, a similar argument leads to the same conclusion. $\square$

Let

$$\psi_0 = \frac{\varphi_0}{\|\varphi_0\|},$$

$$G_1(x,\xi) = G(x,\xi) - \mu_0\psi_0(x)\bar{\psi}_0(\xi),$$

$$(T_1 u)(x) = \int_a^b G_1(x,\xi)u(\xi)d\xi$$

$$= Tu(x) - \mu_0 \langle u, \psi_0\rangle \psi_0(x) \quad \text{for all } u \in C([a,b]). \tag{2.47}$$

The function G_1 has the same regularity and symmetry properties as G, hence
Lemma 2.23 and Theorem 2.24 apply to the self-adjoint operator T_1. If $\|T_1\| \neq 0$
then we define

$$\sup\{|\langle T_1 u, u\rangle| : u \in C([a,b]), \|u\| = 1\} = |\mu_1|,$$

where μ_1 is a (non-zero) real number which, by Theorem 2.24, is an eigenvalue of
T_1 corresponding to some (non-zero) eigenfunction $\varphi_1 \in C([a,b])$, that is,

$$T_1\varphi_1 = \mu_1\varphi_1.$$

Let $\psi_1 = \varphi_1/\|\varphi_1\|$ Then, for any $u \in C([a,b])$,

$$\langle T_1 u, \psi_0\rangle = \langle Tu, \psi_0\rangle - \mu_0\langle\langle u, \psi_0\rangle\psi_0, \psi_0\rangle$$
$$= \langle u, T\psi_0\rangle - \mu_0\langle u, \psi_0\rangle$$
$$= 0$$

since $T\psi_0 = \mu_0\psi_0$. In particular, $\langle T_1\psi_1, \psi_0\rangle = \langle \mu_1\psi_1, \psi_0\rangle = 0$, and therefore
ψ_1 is orthogonal to ψ_0. Now Eq. (2.47) gives

$$T\psi_1 = T_1\psi_1 = \mu_1\psi_1.$$

Thus ψ_1 is also an eigenfunction of T, and the associated eigenvalue satisfies

$$|\mu_1| = \|T\psi_1\| \leq \|T\| = |\mu_0|.$$

Setting

$$G_2(x,\xi) = G_1(x,\xi) - \mu_1\psi_1(x)\bar{\psi}_1(\xi) = G(x,\xi) - \sum_{k=0}^{1}\mu_k\psi_k(x)\bar{\psi}_k(\xi)$$

$$T_2 u = T_1 u - \mu_1 \langle u, \psi_1 \rangle \psi_1 = Tu - \sum_{k=0}^{1} \mu_k \langle u, \psi_k \rangle \psi_k,$$

and proceeding as above, we deduce the existence of a third normalized eigenfunction ψ_2 of T associated with a real eigenvalue μ_2 such that ψ_2 is orthogonal to both ψ_0 and ψ_1, and $|\mu_2| \leq |\mu_1|$. Thus we obtain an orthonormal sequence of eigenfunctions $\psi_0, \psi_1, \psi_2, \cdots$ of T corresponding to the sequence of eigenvalues $|\mu_0| \geq |\mu_1| \geq |\mu_2| \geq \cdots$. The sequence of eigenfunctions terminates only if $\|T_n\| = 0$ for some n. In that case

$$0 = LT_n u = LTu - \sum_{k=0}^{n-1} \mu_k \langle u, \psi_k \rangle L\psi_k = u - \sum_{k=0}^{n-1} \mu_k \langle u, \psi_k \rangle L\psi_k$$

for all $u \in C([a, b])$, which would imply

$$u = \sum_{k=0}^{n-1} \mu_k \langle u, \psi_k \rangle L\psi_k = \sum_{k=0}^{n-1} \langle u, \psi_k \rangle LT\psi_k = \sum_{k=0}^{n-1} \langle u, \psi_k \rangle \psi_k.$$

But because no finite set of eigenfunctions can span $C([a, b])$, this last equality cannot hold for all $u \in C([a, b])$. Consequently, $\|T_n\| > 0$ for all $n \in \mathbb{N}$, and we have the following result.

Theorem 2.25 (Existence Theorem) *The integral operator T defined by Eq. (2.41) has an infinite sequence of eigenfunctions (ψ_n) which are orthonormal in $\mathcal{L}^2(a, b)$.*

2.4.2 Completeness of the Eigenfunctions

If f is any function in $\mathcal{L}^2(a, b)$ then, by Bessel's inequality (1.24), we have

$$\sum_{k=0}^{\infty} |\langle f, \psi_k \rangle|^2 \leq \|f\|^2,$$

where (ψ_n) is the sequence of orthonormal eigenfunctions of T. To prove the completeness of these eigenfunctions in $\mathcal{L}^2(a, b)$ we have to show that this inequality is, in fact, an equality. This we do by first proving that

$$f = \sum_{k=0}^{\infty} \langle f, \psi_k \rangle \psi_k$$

for any $f \in C^2([a, b])$ which satisfies the boundary conditions (2.35) and (2.36), and then using the density of $C^2([a, b])$ in $\mathcal{L}^2(a, b)$ to extend this equality to $\mathcal{L}^2(a, b)$.

Theorem 2.26 *Given any function* $f \in C^2([a, b])$ *which satisfies the boundary conditions* (2.35) *and* (2.36), *the infinite series* $\sum \langle f, \psi_k \rangle \psi_k$ *converges uniformly to* f *on* $[a, b]$.

Proof For every fixed $x \in [a, b]$, we have

$$\langle G(x, \cdot), \psi_k \rangle = T\bar{\psi}_k(x) = \mu_k \bar{\psi}_k(x) \quad \text{for all } k \in \mathbb{N}.$$

Bessel's inequality, applied to G as a function of ξ, yields

$$\sum_{k=0}^{n} \mu_k^2 \left| \psi_k(x) \right|^2 \le \int_a^b |G(x, \xi)|^2 \, d\xi \quad \text{for all } x \in [a, b], \ n \in \mathbb{N}.$$

Integrating with respect to x and letting $n \to \infty$, we obtain

$$\sum_{k=0}^{\infty} \mu_k^2 \le M^2 (b - a)^2, \tag{2.48}$$

where M is the maximum value of $|G(x, \xi)|$ on the square $[a, b] \times [a, b]$. An immediate consequence of the inequality (2.48) is that

$$\lim_{n \to \infty} \left| \mu_n \right| = 0. \tag{2.49}$$

With

$$G_n(x, \xi) = G(x, \xi) - \sum_{k=0}^{n-1} \mu_k \psi_k(x) \bar{\psi}_k(\xi),$$

we have already seen that the integral operator

$$T_n : u(x) \mapsto \int_a^b G_n(x, \xi) u(\xi) d\xi, \ u \in C([a, b]),$$

has eigenvalue μ_n and norm

$$\|T_n\| = \left| \mu_n \right|.$$

Therefore, for any $u \in C([a, b])$,

$$\|T_n u\| = \left\| T u - \sum_{k=0}^{n-1} \mu_k \langle u, \psi_k \rangle \psi_k \right\| \leq |\mu_n| \, \|u\| \to 0 \ \text{ as } n \to \infty \tag{2.50}$$

in view of (2.49). If $n > m$, then

$$\sum_{k=m}^{n} \mu_k \langle u, \psi_k \rangle \psi_k = T \left(\sum_{k=m}^{n} \langle u, \psi_k \rangle \psi_k \right) ;$$

but since $|T u| \leq M \sqrt{b - a} \, \|u\|$ for all $u \in C([a, b])$, it follows that

$$\left| \sum_{k=m}^{n} \mu_k \langle u, \psi_k \rangle \psi_k \right| \leq \|T\| \left\| \sum_{k=m}^{n} \langle u, \psi_k \rangle \psi_k \right\|$$

$$\leq M \sqrt{b - a} \left(\sum_{k=m}^{n} |\langle u, \psi_k \rangle|^2 \right)^{1/2} .$$

By Bessel's inequality, the right-hand side of this inequality tends to zero as m, $n \to \infty$, hence the series

$$\sum_{k=0}^{\infty} \mu_k \langle u, \psi_k \rangle \psi_k$$

converges uniformly on $[a, b]$ to a continuous function. The continuity of $T u$ and (2.50) now imply

$$T u(x) = \sum_{k=0}^{\infty} \mu_k \langle u, \psi_k \rangle \psi_k(x) \ \text{ for all } x \in [a, b]. \tag{2.51}$$

If $f \in C^2([a, b])$ satisfies the boundary conditions (2.35) and (2.36), then $u = Lf$ is a continuous function on $[a, b]$ and $f = T u$. With

$$\mu_k \langle u, \psi_k \rangle = \langle u, \mu_k \psi_k \rangle = \langle u, T \psi_k \rangle = \langle T u, \psi_k \rangle = \langle f, \psi_k \rangle ,$$

Eq. (2.51) becomes

$$f(x) = \sum_{k=0}^{\infty} \langle f, \psi_k \rangle \psi_k(x) \ \text{ for all } x \in [a, b]. \tag{2.52}$$

$\square$

To prove the completeness of the orthonormal sequence $(\psi_n : n \in \mathbb{N}_0)$ of eigenfunctions of T in $\mathcal{L}^2(a, b)$, it remains to show that Eq. (2.52) holds in $\mathcal{L}^2$ for *any* $f \in \mathcal{L}^2(a, b)$. Since equality in $\mathcal{L}^2$ is not affected by the values of f or f' at the end points a and b, it suffices to show that $C^2([a, b])$ is dense in $\mathcal{L}^2(a, b)$. But this follows from the density of $C^2([a, b])$ in $C([a, b])$ as a direct consequence of the Weierstrass approximation theorem (see [1]), and the density of $C([a, b])$ in $\mathcal{L}^2(a, b)$ as stated in Theorem 1.27. Hence, if $f \in \mathcal{L}^2(a, b)$, then the seqence of C^2 functions $\sum_{k=0}^{n} \langle f, \psi_k \rangle \psi_k$ approaches f in $\mathcal{L}^2$ as $n \to \infty$, or

$$\lim_{n \to \infty} \left\| f - \sum_{k=0}^{n} \langle f, \psi_k \rangle \psi_k \right\| = 0. \tag{2.53}$$

Equation (2.53) is equivalent to the equation

$$f \overset{\mathcal{L}^2}{=} \sum_{k=0}^{\infty} \langle f, \psi_k \rangle \psi_k.$$

It is also equivalent to Parseval's relation

$$\| f \|^2 = \sum_{k=0}^{\infty} |\langle f, \psi_k \rangle|^2 .$$

Each one these equations expresses the fact that the sequence (ψ_k) of orthonormal eigenfunctions of T forms a complete set in $\mathcal{L}^2(a, b)$, and constitutes an orthonormal basis for the space.

Going back to the SL problem

$$Lu + \lambda u = 0,$$

$$\alpha_1 u(a) + \alpha_2 u'(a) = 0,$$

$$\beta_1 u(b) + \beta_2 u'(b) = 0,$$

where $u \in C^2([a, b])$, we have shown that this system of equations is equivalent to the single integral equation

$$Tu = \mu u = -u/\lambda.$$

Since the boundary conditions determine a unique solution to the differential equation $Lu + \lambda u = 0$, each eigenvalue λ of $-L$ corresponds to a unique eigenfunction u (up to a multiplicative constant). In this case each eigenvalue is said to have *multiplicity one,* or *simple multiplicity.*

By Eq. (2.49), $1/|\lambda_n| = |\mu_n| \to 0$, and it follows from Lemma 2.21 that $\lambda_n \to \infty$ as $n \to \infty$. Reintroducing the weight function ρ, we therefore arrive at the following fundamental theorem of the completeness of the eigenfunctions of the regular SL problem in $\mathcal{L}^2$.

Theorem 2.27 (Completenes of the Eigenfunctions) *Assuming* $p', r, \rho \in C([a, b])$, *and* $p, \rho > 0$ *on* $[a, b]$, *the SL eigenvalue problem defined by Eqs. (2.34)–(2.36) has an infinite sequence of real eigenvalues*

$$\lambda_0 < \lambda_1 < \lambda_2 < \cdots$$

such that $\lambda_n \to \infty$. *To each eigenvalue* λ_n *corresponds a single eigenfunction* φ_n, *and the sequence of eigenfunctions* $(\varphi_n : n \in \mathbb{N}_0)$ *forms an orthogonal basis of* $\mathcal{L}^2_\rho(a, b)$.

Remark 2.28

1. If the separated boundary conditions (2.35) and (2.36) are replaced by the periodic conditions

$$u(a) = u(b), \ u'(a) = u'(b),$$

   Eq. (2.26) is satisfied provided $p(a) = p(b)$ (Exercise 2.26). The operator defined by (2.33) will then be self-adjoint and its eigenvalues will be bounded below by Lemma 2.21. The results obtained above remain valid, except that the uniqueness of the eigenfunction for each eigenvalue is not guaranteed. This is illustrated by Example 3.1 in the next chapter, where each nonzero eigenvalue has two linearly independent eigenfunctions. Of course we cannot have more than two linearly independent eigenfunctions, since the differential equation is of second order. This would seem to imply that periodic boundary values are less restrictive than seperated boundary conditions. For most real values of λ, however, i.e those which are not eigenvalues, the boundary conditions are already too restrictive to admit any (non-trivial) eigenfunction.

2. It can also be shown that each eigenfunction φ_n, corresponding to the eigenvalue λ_n, has exactly n zeros in the interval (a, b), and it is worth checking this claim against the examples which will be discussed henceforth. A proof of this result may be found in [6]. More on the SL problem and the distribution of its eigenvalues may be found in [3].

Example 2.29 Find the eigenvalues and eigenfunctions of the equation

$$u'' + \lambda u = 0, \ 0 \le x \le l,$$

subject to one of the pairs of separated boundary conditions

$$(i) \ u(0) = u(l) = 0,$$
$$(ii) \ u'(0) = u'(l) = 0.$$

Solution The differential operator is formally self-adjoint and the boundary conditions (i) and (ii) are homogeneous and separated, so both Theorems 2.14 and 2.27 apply. We have already seen in Example 2.15 that, under the boundary conditions (i), the equation $u'' + \lambda u = 0$ has only positive eigenvalues. Under the boundary conditions (ii) it has nonnegative eigenvalues.

If $\lambda > 0$, the general solution of the differential equation is

$$u(x) = c_1 \cos \sqrt{\lambda}x + c_2 \sin \sqrt{\lambda}x. \tag{2.54}$$

Applying conditions (i), we obtain

$$u(0) = c_1 = 0,$$

$$u(l) = c_2 \sin \sqrt{\lambda}l = 0 \Rightarrow \lambda_n = \frac{n^2\pi^2}{l^2}, \quad n \in \mathbb{N}.$$

In accordance with Theorem 2.27 the eigenvalues λ_n tend to ∞, and the corresponding eigenfunctions

$$u_n(x) = \sin \frac{n\pi}{l}x$$

are orthogonal in $\mathcal{L}^2(0, l)$, since, for all $m \neq n$,

$$\int_0^l \sin \frac{m\pi}{l}x \sin \frac{n\pi}{l}x \, dx = \frac{1}{2} \int_0^l \left[\cos(m - n)\frac{\pi}{l}x - \cos(m + n)\frac{\pi}{l}x \right] dx = 0.$$

From Theorem 2.28 we also conclude that the sequence of functions

$$\sin \frac{n\pi}{l}x, \quad n \in \mathbb{N},$$

is complete in $\mathcal{L}^2(0, l)$.

Now we turn to the boundary conditions (ii): If $\lambda = 0$, the general solution of $u'' = 0$ is

$$u(x) = c_1 x + c_2,$$

and conditions (ii) imply $c_1 = 0$. u is therefore a constant, which we can take to be 1. Thus

$$\lambda_0 = 0, \ u_0(x) = 1$$

is the first eigenvalue-eigenfunction pair. But if $\lambda > 0$, the first of conditions (ii) applied to (2.54) gives

$$u'(0) = \sqrt{\lambda}c_2 = 0,$$

hence $c_2 = 0$. The condition at $x = l$ yields

$$u'(l) = -c_1\sqrt{\lambda}\sin\sqrt{\lambda}l = 0 \Rightarrow \lambda_n = \frac{n^2\pi^2}{l^2}$$

$$u_n(x) = \cos\frac{n\pi}{l}x, \quad n \in \mathbb{N}.$$

Thus the eigenvalues and eigenfunctions under conditions (ii) are

$$\lambda_n = \frac{n^2\pi^2}{l^2}, \quad u_n(x) = \cos\frac{n\pi}{l}x, \quad n \in \mathbb{N}_0.$$

Again we can verify that the sequence $\left(\cos\dfrac{n\pi}{l}x : n \in \mathbb{N}_0\right)$ is orthogonal in $\mathcal{L}^2(0, l)$, and we conclude from Theorem 2.28 that it also spans this space.

According to Example 2.29, any function $f \in \mathcal{L}^2(0, l)$ can be represented by an infinite series of the form

$$f(x) = \sum_{n=0}^{\infty} a_n \cos\frac{n\pi}{l}x, \tag{2.55}$$

or of the form

$$f(x) = \sum_{n=1}^{\infty} b_n \sin\frac{n\pi}{l}x, \tag{2.56}$$

the equality in each case being in $\mathcal{L}^2(0, l)$. Note that these two representations of $f(x)$ do not necessarily give the same value at *every* point in $[0, l]$. For example, at $x = 0$, we obtain $f(0) = \sum_{n=0}^{\infty} a_n$ in the first representation and $f(0) = 0$ in the second. That is because Eqs. (2.55) and (2.56) are not pointwise but $\mathcal{L}^2(0, l)$ equalities, and should therefore be interpreted to mean

$$\left\| f(x) - \sum_{n=0}^{\infty} a_n \cos\frac{n\pi}{l}x \right\| = 0, \quad \left\| f(x) - \sum_{n=1}^{\infty} b_n \sin\frac{n\pi}{l}x \right\| = 0,$$

respectively.

By Definition 1.29, the coefficients a_n and b_n are determined by the formulas

$$a_n = \frac{\langle f, \cos(n\pi x/l)\rangle}{\|\cos(n\pi x/l)\|^2}, \quad n \in \mathbb{N}_0, \quad b_n = \frac{\langle f, \sin(n\pi x/l)\rangle}{\|\sin(n\pi x/l)\|^2}, \quad n \in \mathbb{N}.$$

Since

$$\left\|\cos\left(\frac{n\pi}{l}x\right)\right\|^2 = \int_0^l \cos^2\left(\frac{n\pi}{l}x\right) dx = \begin{cases} l, & n = 0 \\ l/2, & n \in \mathbb{N} \end{cases},$$

$$\left\|\sin\left(\frac{n\pi}{l}x\right)\right\|^2 = \int_0^l \sin^2\left(\frac{n\pi}{l}x\right) dx = l/2, \ n \in \mathbb{N},$$

the coefficients in the expansions (2.55) and (2.56) are therefore given by

$$a_0 = \frac{1}{l}\int_0^l f(x)dx, \tag{2.57}$$

$$a_n = \frac{2}{l}\int_0^l f(x)\cos\frac{n\pi}{l}x \, dx, \tag{2.58}$$

$$b_n = \frac{2}{l}\int_0^l f(x)\sin\frac{n\pi}{l}x \, dx. \tag{2.59}$$

Consider, for example, the constant function defined on $[0, l]$ by $f(x) = 1$. Since, for every $n \in \mathbb{N}$,

$$\langle 1, \sin(n\pi x/l)\rangle = \int_0^l \sin\frac{n\pi x}{l}dx = \frac{l}{n\pi}(1 - \cos n\pi) = \frac{l}{n\pi}(1 - (-1)^n),$$

we have $b_n = \dfrac{2}{n\pi}(1 - (-1)^n)$, so we can write

$$1 = \frac{2}{\pi}\sum_{n=1}^{\infty}\left(\frac{1 - (-1)^n}{n}\right)\sin\frac{n\pi}{l}x$$

$$= \frac{4}{\pi}\left(\sin\frac{\pi}{l}x + \frac{1}{3}\sin\frac{3\pi}{l}x + \frac{1}{5}\sin\frac{5\pi}{l}x + \cdots\right) \ \text{in } \mathcal{L}^2\,(0, l),$$

or

$$\frac{2}{\pi}\sum_{k=1}^{n}\left(\frac{1 - (-1)^k}{k}\right)\sin\frac{k\pi}{l}x \xrightarrow{\mathcal{L}^2} 1 \ \text{as } n \to \infty.$$

On the other hand,

$$\langle 1, \cos(n\pi x/l)\rangle = \int_0^l \cos(n\pi x/l)dx = \begin{cases} l, & n = 0 \\ 0, & n \in \mathbb{N} \end{cases},$$

$$\|1\|^2 = l,$$

so the coefficients a_n in the expansion (2.55) all vanish except for the first, which is $a_0 = 1$, hence the series reduces to the single term 1. That is because the function $f = 1$ coincides with the first element of the sequence $\left(\cos \frac{n\pi x}{l} : n \in \mathbb{N}_0\right)$.

By the same token, based on Theorem 2.27, the sequence of eigenfunctions for the SL problem discussed in Example 2.20 is complete in $\mathcal{L}_\rho^2(1, b)$, where $\rho(x) = 1/x$. Hence any function $f \in \mathcal{L}_\rho^2(1, b)$ can be expanded in a series of the form

$$\sum_{n=1}^{\infty} b_n \sin\left(\frac{n\pi}{\log b} \log x\right),$$

where

$$b_n = \frac{\langle f, \sin(n\pi \log x / \log b)\rangle_\rho}{\|\sin(n\pi \log x / \log b)\|_\rho^2}.$$

2.4.3   The Singular SL Problem

We end this section with a brief word on the singular SL problem. In the equation

$$Lu + \lambda\rho u = (pu')' + ru + \lambda\rho u = 0, \quad a < x < b,$$

where p is smooth and ρ is positive and continuous, we have so far assumed that p does not vanish on the closed and bounded interval $[a, b]$. Any relaxation of these conditions leads to a singular problem. In this book we consider singular SL problems which result from one or both of the following cases:

1. $p(x) = 0$ at $x = a$ and/or $x = b$.
2. the interval (a, b) is infinite.

In the first instance the expression $\rho p(f'g - fg')$ vanishes at the end-point where $p = 0$, hence no boundary condition is required at that end-point. In particular, if $p(a) = p(b) = 0$ and the limits of ρu at a and b exist, then the equation

$$\rho p(f'g - fg')\big|_a^b = 0 \tag{2.60}$$

is satisfied and L is self-adjoint. If (a, b) is infinite, the inner product is treated as an improper integral, and here the weight function ρ often plays a role in securing its convergence as $|x| \to \infty$. The conclusions of Theorem 2.27 remain valid under these conditions, as the examples in chapters 4 and 5 demonstrate, though we do not prove that. But the interested reader may wish to look up some titles in the reference list, such as [16], [18], and [20], for more on this subject.

In the chapters to follow we consider the following typical examples of singular SL equations.

1. *Legendre's Equation*

$$(1 - x^2)u'' - 2xu' + n(n + 1)u = 0, \quad -1 < x < 1,$$

where $\lambda = n(n + 1)$ and the function $p(x) = 1 - x^2$ vanishes at the end-points $x = \pm 1$.

2. *Hermite's Equation*

$$u'' - 2xu' + 2nu = 0, \quad x \in \mathbb{R}.$$

After multiplication by e^{-x^2}, this equation is transformed to the standard self-adjoint form

$$e^{-x^2}u'' - 2xe^{-x^2}u' + 2ne^{-x^2}u = 0,$$

where $\lambda = 2n$, $p(x) = e^{-x^2}$, and $\rho(x) = e^{-x^2}$.

3. *Laguerre's Equation*

$$xu'' - (1 - x)u' + nu = 0, \quad x > 0,$$

which is transformed to the standard form

$$xe^{-x}u'' - (1 - x)e^{-x}u' + ne^{-x}u = 0$$

by multiplication by e^{-x}. Here $p(x) = xe^{-x}$ vanishes at $x = 0$.

4. *Bessel's Equation*

$$xu'' + u' - \frac{n^2}{x}u + \lambda xu = 0, \quad x > 0.$$

Here both functions $p(x) = x$ and $\rho(x) = x$ vanish at $x = 0$.

The solutions of these equations provide important examples of the so-called special functions of mathematical physics, which make up the subject of Chaps. 4 and 5. But before that, in Chap. 3, we use the regular SL problem to develop the $\mathcal{L}^2$ theory of Fourier series. The singular problems discussed in the following chapters demonstrate how it can be extended, and the result is surprisingly rich in theory and very useful in application.

Exercises

2.23 Determine the eigenvalues and eigenfunctions of the boundary-value problem

$$u'' + \lambda u = 0, \quad a \le x \le b,$$
$$u(a) = u(b) = 0.$$

2.24 Determine the eigenvalues and eigenfunctions of the equation $x^2 u'' - xu' + \lambda u = 0$ on $(1, e)$ under the boundary conditions $u(1) = u(e) = 0$. Write the form of the orthogonality relation between the eigenfunctions.

2.25 If f and g satisfy the separated homogeneous boundary conditions (2.35) and (2.36), verify that $\left. p(f'\bar{g} - f\bar{g}') \right|_a^b = 0$

2.26 Verify that $\left. p(f'\bar{g} - f\bar{g}') \right|_a^b = 0$ if f and g satisfy the periodic conditions $u(a) = u(b), \; u'(a) = u'(b)$, provided $p(a) = p(b)$.

2.27 Which of the following boundary conditions make $\left. p(f'g - fg') \right|_a^b = 0$?

(a) $p(x) = 1, \; a \le x \le b, \; u(a) = u(b), \; u'(a) = u'(b)$.
(b) $p(x) = x, \; 0 < a \le x \le b, \; u(a) = u'(b) = 0$.
(c) $p(x) = \sin x, \; 0 \le x \le \pi/2, \; u(0) = 1, \; u(\pi/2) = 0$.
(d) $p(x) = e^{-x}, \; 0 < x < 1, \; u(0) = u(1), \; u'(0) = u'(1)$.
(e) $p(x) = x^2, \; 0 < x < b, \; u(0) = u'(b), \; u(b) = u'(0)$.
(f) $p(x) = x^2, \; -1 < x < 1, \; u(-1) = u(1), \; u'(-1) = u'(1)$.

2.28 Which boundary conditions in Exercise 2.27 define a singular SL problem?

2.29 Determine the eigenvalues and eigenfunctions of the problem

$$[(x+3)^2 y']' + \lambda y = 0, \quad -2 \le x \le 1,$$
$$y(-2) = y(1) = 0.$$

2.30 Let

$$u'' + \lambda u = 0, \quad 0 \le x \le \pi,$$
$$u(0) = 0, \quad 2u(\pi) - u'(\pi) = 0.$$

(a) Show how the positive eigenvalues λ may be determined. Are there any eigenvalues in $(-\infty, 0]$?
(b) What are the corresponding eigenfunctions?

2.31 Discuss the sequence of eigenvalues and eigenfunctions of the problem

$$u'' + \lambda u = 0, \quad 0 \le x \le l,$$
$$u'(0) = u(0), \quad u'(l) = 0.$$

2.32 Let

$$(pu')' + ru + \lambda u = 0, \quad a < x < b,$$

$$u(a) = u(b) = 0.$$

If $p(x) \geq 0$ and $r(x) \leq c$, prove that $\lambda \geq -c$.

2.33 Prove that the norm of the operator T, as defined in Eq. (2.42), is also given by

$$\|T\| = \sup \left\{ |\langle Tu, u \rangle| : u \in C\left([a, b]\right), \|u\| = 1 \right\}.$$

Hint: $|\langle Tu, u \rangle| \leq \|T\|$ by the CBS inequality. For the reverse inequality show that $2\operatorname{Re}\langle Tu, v \rangle \leq \left(\|u\|^2 + \|v\|^2 \right) \sup |\langle Tu, u \rangle|$.

Chapter 3
Fourier Series

This chapter deals with the theory and applications of *Fourier series*, named after *Joseph Fourier* (1768–1830), the French physicist who introduced the series in his investigation of the transfer of heat. His results were later refined by others, especially the German mathematician *Gustav Lejeune Dirichlet* (1805–1859), who made important contributions to the convergence properties of the series.

The first section presents the $\mathcal{L}^2$ theory, which is based on the results of the regular SL problem as developed in the previous chapter. We saw in Example 2.29 that, based on Theorem 2.27, both sets of orthogonal functions $\{\cos(n\pi x/l) : n \in \mathbb{N}_0\}$ and $\{\sin(n\pi x/l) : n \in \mathbb{N}\}$ span the space $\mathcal{L}^2(0, l)$. We now show that their union spans $\mathcal{L}^2(-l, l)$, and thereby arrive at the fundamental theorem of Fourier series.

In Sect. 3.2 we consider the pointwise theory and prove the basic result in this connection, which is Theorem 3.9, using the properties of the Dirichlet kernel. The third and final section of this chapter is devoted to applications to boundary-value problems.

3.1 Fourier Series in $\mathcal{L}^2$

A function $\varphi : [-l, l] \to \mathbb{C}$ is said to be even if

$$\varphi(-x) = \varphi(x) \ \text{ for all } x \in [-l, l],$$

and odd if

$$\varphi(-x) = -\varphi(x) \ \text{ for all } x \in [-l, l].$$

© The Author(s), under exclusive license to Springer-Verlag London Ltd., part of Springer Nature 2026
M. A. Al-Gwaiz, *Sturm-Liouville Theory and its Applications*, Springer Undergraduate Mathematics Series, https://doi.org/10.1007/978-1-4471-7610-7_3

The same definition applies if the compact interval $[-l, l]$ is replaced by $(-l, l)$ or $(-\infty, \infty)$. If φ is integrable on $[-l, l]$, then it follows that

$$\int_{-l}^{l} \varphi(x)dx = 2\int_{0}^{l} \varphi(x)dx$$

if φ is even, and

$$\int_{-l}^{l} \varphi(x)dx = 0$$

if φ is odd.

An arbitrary function f defined on $[-l, l]$ can always be represented as a sum of the form

$$f(x) = \frac{1}{2}[f(x) + f(-x)] + \frac{1}{2}[f(x) - f(-x)], \tag{3.1}$$

where $f_e(x) = \frac{1}{2}[f(x) + f(-x)]$ is an even function and $f_o(x) = \frac{1}{2}[f(x) - f(-x)]$ is odd. f_e and f_o are referred to as the even and the odd components of f, respectively. To show that the representation (3.1) is unique, let

$$f(x) = \tilde{f}_e(x) + \tilde{f}_o(x) = f_e(x) + f_o(x), \tag{3.2}$$

where $\tilde{f}_e$ is even and $\tilde{f}_o$ is odd. Replacing x by $-x$ gives

$$f(-x) = \tilde{f}_e(x) - \tilde{f}_o(x) = f_e(x) - f_o(x). \tag{3.3}$$

Adding and subtracting (3.2) and (3.3) we conclude that $\tilde{f}_e = f_e$ and $\tilde{f}_o = f_o$.

Since any linear combination of even (odd) functions is even (odd), the even and odd functions in $\mathcal{L}^2(-l, l)$ clearly form two complementary subspaces which are orthogonal to each other, in the sense that if $\varphi, \psi \in \mathcal{L}^2(-l, l)$ with φ even and ψ odd, then φ and ψ are orthogonal to each other, for their inner product is

$$\langle \varphi, \psi \rangle = \int_{-l}^{l} \varphi(x)\psi(x)dx = 0$$

because the function $\varphi\psi$ is odd.

Let $f = f_e + f_o$ be any function in $\mathcal{L}^2(-l, l)$. Clearly both f_e and f_o belong to $\mathcal{L}^2(-l, l)$. The restriction of f_e to $(0, l)$ can be represented in $\mathcal{L}^2$ in terms of the (complete) orthogonal set $\{\cos(n\pi x/l) : n \in \mathbb{N}_0\}$ as

$$f_e(x) = a_0 + \sum_{n=1}^{\infty} a_n \cos\frac{n\pi}{l}x, \quad 0 \le x \le l, \tag{3.4}$$

where, according to the formulas (2.56) and (2.57),

$$a_0 = \frac{1}{l} \int_0^l f_e(x)dx,$$

$$a_n = \frac{2}{l} \int_0^l f_e(x) \cos \frac{n\pi}{l}x \, dx, \quad n \in \mathbb{N}.$$

The orthogonal set $\{\sin(n\pi x/l) : n \in \mathbb{N}\}$ also spans $\mathcal{L}^2(0, l)$, hence

$$f_o(x) = \sum_{n=1}^{\infty} b_n \sin \frac{n\pi}{l}x, \quad 0 \le x \le l, \tag{3.5}$$

where

$$b_n = \frac{2}{l} \int_0^l f_o(x) \sin \frac{n\pi}{l}x \, dx, \quad n \in \mathbb{N}.$$

Since both sides of (3.4) are even functions on $[-l, l]$, the equality extends to $[-l, l]$; and similarly the two sides of Eq. (3.5) are odd and this allows us to extend the equality to $[-l, l]$. Hence we have the following representation of any $f \in \mathcal{L}^2(-l, l)$,

$$f(x) = \left(a_0 + \sum_{n=1}^{\infty} a_n \cos \frac{n\pi}{l}x\right) + \sum_{n=1}^{\infty} b_n \sin \frac{n\pi}{l}x$$

$$= a_0 + \sum_{n=1}^{\infty} \left(a_n \cos \frac{n\pi}{l} + b_n \sin \frac{n\pi}{l}x\right), \quad -l \le x \le l. \tag{3.6}$$

Furthermore, using the properties of even and odd functions, we have, for all $n \in \mathbb{N}_0$,

$$\int_{-l}^{l} f_e(x) \sin \frac{n\pi}{l}x \, dx = \int_{-l}^{l} f_o(x) \cos \frac{n\pi}{l}x \, dx = 0,$$

$$\int_0^l f_e(x) \cos \frac{n\pi}{l}x \, dx = \frac{1}{2} \int_{-l}^{l} f_e(x) \cos \frac{n\pi}{l}x \, dx = \frac{1}{2} \int_{-l}^{l} f(x) \cos \frac{n\pi}{l}x \, dx,$$

$$\int_0^l f_o(x) \sin \frac{n\pi}{l}x \, dx = \frac{1}{2} \int_{-l}^{l} f_o(x) \sin \frac{n\pi}{l}x \, dx = \frac{1}{2} \int_{-l}^{l} f(x) \sin \frac{n\pi}{l}x \, dx.$$

Hence the coefficients a_n and b_n in the representation (3.6) are given by

$$a_0 = \frac{1}{2l} \int_{-l}^{l} f(x)\,dx,$$

$$a_n = \frac{1}{l} \int_{-l}^{l} f(x) \cos \frac{n\pi}{l} x\, dx, \quad n \in \mathbb{N},$$

$$b_n = \frac{1}{l} \int_{-l}^{l} f(x) \sin \frac{n\pi}{l} x\, dx, \quad n \in \mathbb{N}.$$

This result is illustrated by the following example.

Example 3.1 Find the eigenfunctions of the equation

$$u'' + \lambda u = 0, \quad -l \le x \le l,$$

which satisfy the periodic boundary conditions

$$u(-l) = u(l), \quad u'(-l) = u'(l).$$

Solution This is an SL problem in which the boundary conditions are not separated but periodic. Since the operator $-\dfrac{d^2}{dx^2}$ is self-adjoint under these conditions (see Exercise 2.26), the conclusions of Theorem 2.27 apply except for the uniqueness of the eigenfunctions, as pointed out in Remark 2.28, hence its eigenfunctions are orthogonal and complete in $\mathcal{L}^2(-l, l)$. If $\lambda < 0$ the differential equation has only the trivial solution under the given periodic boundary conditions. If $\lambda = 0$ the general solution is $u(x) = c_1 x + c_2$, and the boundary conditions yield the eigenfunction $u_0(x) = 1$.

For positive values of λ the solution of the equation is

$$c_1 \cos \sqrt{\lambda} x + c_2 \sin \sqrt{\lambda} x.$$

Applying the boundary conditions to this solution leads to the pair of equations

$$c_2 \sin \sqrt{\lambda} l = 0,$$

$$c_1 \sqrt{\lambda} \sin \sqrt{\lambda} l = 0.$$

Since $\sqrt{\lambda} > 0$ and c_1 and c_2 cannot both be 0, the eigenvalues must satisfy $\sin \sqrt{\lambda}\, l = 0$. Thus the eigenvalues in $[0, \infty)$ are given by

$$\lambda_n = \left(\frac{n\pi}{l}\right)^2, \quad n \in \mathbb{N}_0,$$

and the corresponding eigenfunctions are

$$1, \ \cos\frac{\pi}{l}x, \ \sin\frac{\pi}{l}x, \ \cos\frac{2\pi}{l}x, \ \sin\frac{2\pi}{l}x, \ \cos\frac{3\pi}{l}x, \ \sin\frac{3\pi}{l}x, \ \cdots.$$

Note that every eigenvalue $\lambda_n = (n\pi/l)^2$ is associated with two eigenfunctions, $\cos(n\pi x/l)$ and $\sin(n\pi x/l)$, except $\lambda_0 = 0$ which is associated with the single eigenfunction 1.

We have therefore proved the following result.

Theorem 3.2 (Fundamental Theorem of Fourier Series) *The orthogonal set of functions*

$$\{1, \cos\frac{n\pi}{l}x, \sin\frac{n\pi}{l}x : n \in \mathbb{N}\}$$

is complete in $\mathcal{L}^2(-l, l)$, *in the sense that any function* $f \in \mathcal{L}^2(-l, l)$ *can be represented by the series*

$$f(x) = a_0 + \sum_{n=1}^{\infty}\left(a_n \cos\frac{n\pi}{l}x + b_n \sin\frac{n\pi}{l}x\right), \tag{3.7}$$

where the equality is in $\mathcal{L}^2(-l, l)$, *and*

$$a_0 = \frac{\langle f, 1\rangle}{\|1\|^2} = \frac{1}{2l}\int_{-l}^{l} f(x)dx, \tag{3.8}$$

$$a_n = \frac{\langle f, \cos(n\pi x/l)\rangle}{\|\cos(n\pi x/l)\|^2} = \frac{1}{l}\int_{-l}^{l} f(x)\cos\frac{n\pi}{l}x \, dx, \ n \in \mathbb{N}, \tag{3.9}$$

$$b_n = \frac{\langle f, \sin(n\pi x/l)\rangle}{\|\sin(n\pi x/l)\|^2} = \frac{1}{l}\int_{-l}^{l} f(x)\sin\frac{n\pi}{l}x \, dx, \ n \in \mathbb{N}. \tag{3.10}$$

The right-hand side of Eq. (3.7) is called the *Fourier series* expansion of f, and the coefficients a_n and b_n in the expansion are the corresponding *Fourier coefficients*.

Remark 3.3

1. If f is an even function, the coefficients $b_n = 0$ for all $n \in \mathbb{N}$ and f is then represented on $[-l, l]$ by a cosine series,

$$f(x) = a_0 + \sum_{n=1}^{\infty} a_n \cos\frac{n\pi}{l}x,$$

in which the coefficients are given by

$$a_0 = \frac{1}{l} \int_{-l}^{l} f(x)\,dx, \quad a_n = \frac{2}{l} \int_{0}^{l} f(x) \cos \frac{n\pi}{l} x \, dx, \quad n \in \mathbb{N}.$$

Otherwise, when f is odd, $a_n = 0$ for all $n \in \mathbb{N}_0$ and f is represented by a sine series,

$$f(x) = \sum_{n=1}^{\infty} b_n \sin \frac{n\pi}{l} x,$$

where

$$b_n = \frac{2}{l} \int_{0}^{l} f(x) \sin \frac{n\pi}{l} x \, dx, \quad n \in \mathbb{N}.$$

2. The equality expressed by (3.7) between f and its Fourier series is not necessarily pointwise. It means that

$$\left\| f(x) - \left[a_0 + \sum_{n=1}^{N} \left(a_n \cos \frac{n\pi}{l} x + b_n \sin \frac{n\pi}{l} x \right) \right] \right\|^2 =$$

$$\left| \|f\|^2 - l \left[2\,|a_0|^2 + \sum_{n=1}^{N} (|a_n|^2 + |b_n|^2) \right] \right| \to 0 \ \text{ as } N \to \infty.$$

With $f \in \mathcal{L}^2(-l, l)$, this convergence clearly implies that the positive series $\sum |a_n|^2$ and $\sum |b_n|^2$ both converge and therefore $\lim a_n = \lim b_n = 0$.

3. If the Fourier series of f converges uniformly on $[-l, l]$, then, by Corollary 1.16, its sum is continuous and the equality (3.7) is pointwise.

Example 3.4 The function

$$f(x) = \begin{cases} -1, & -\pi < x < 0 \\ 0, & x = 0 \\ 1, & 0 < x \le \pi, \end{cases}$$

shown in Fig. 3.1 is clearly in $\mathcal{L}^2(-\pi, \pi)$. To obtain its Fourier series expansion, we first note that f is odd on $(-\pi, \pi)$, hence its Fourier series reduces to a sine series. The Fourier coefficients are given by

$$b_n = \frac{2}{\pi} \int_{0}^{\pi} \sin nx \, dx$$

$$= \frac{2}{n\pi} (1 - \cos n\pi).$$

Fig. 3.1 The function f

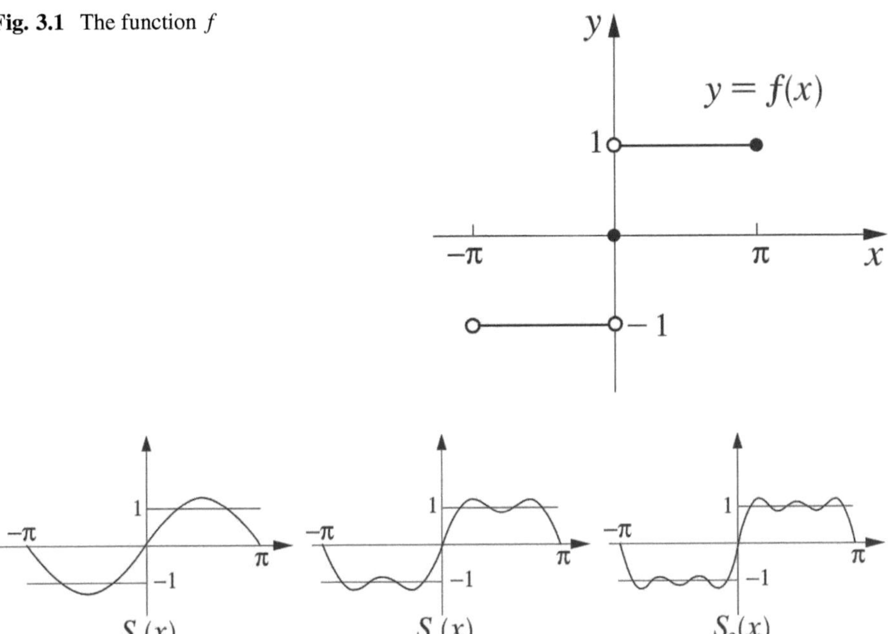

Fig. 3.2 Sequence of partial sums S_N

b_n is therefore $4/n\pi$ when n is odd, and 0 when n is even, and tends to 0 as $n \to \infty$. Thus we have

$$f(x) = \sum_{n=1}^{\infty} b_n \sin nx$$

$$= \frac{4}{\pi} \sin x + \frac{4}{3\pi} \sin 3x + \frac{4}{5\pi} \sin 5x + \cdots$$

$$= \frac{4}{\pi} \sum_{n=0}^{\infty} \frac{1}{2n+1} \sin(2n+1)x. \tag{3.11}$$

Observe how the terms of the partial sums

$$S_N(x) = \frac{4}{\pi} \sum_{n=0}^{N} \frac{1}{2n+1} \sin(2n+1)x$$

add up in Fig. 3.2 to approximate the graph of f. Note also that the Fourier series of f is 0 at $\pm\pi$ whereas $f(\pi) = 1$ and $f(-\pi)$ is not defined. We shall have more to say about that in the next section.

Theorem 3.2 applies to real as well as complex functions in $\mathcal{L}^2(-l, l)$. When f is real there is a clear advantage to using the orthogonal sequence of real trigonometric functions

$$1, \ \cos\frac{\pi}{l}x, \ \sin\frac{\pi}{l}x, \ \cos\frac{2\pi}{l}x, \ \sin\frac{2\pi}{l}x, \ \cdots$$

to expand f, for then the Fourier coefficients are also real. Alternatively, we could use Euler's relation

$$e^{ix} = \cos x + i \sin x, \quad x \in \mathbb{R},$$

to express the Fourier series in the exponential form

$$f(x) = \sum_{-\infty}^{\infty} c_n e^{in\pi x/l}, \quad \text{where} \quad c_n = \frac{\langle f, e^{in\pi x/l}\rangle}{\|e^{in\pi x/l}\|^2}, \ n \in \mathbb{Z}.$$

The coefficients in this case are related to the coefficients in the trigonometric series by

$$c_0 = \frac{\langle f, 1\rangle}{\|1\|^2} = \frac{1}{2l}\int_{-l}^{l} f(x)dx = a_0,$$

$$c_n = \frac{\langle f, e^{in\pi x/l}\rangle}{\|e^{in\pi x/l}\|^2} = \frac{1}{2l}\int_{-l}^{l} f(x)e^{-in\pi x/l}dx = \frac{1}{2}(a_n - ib_n),$$

$$c_{-n} = \frac{\langle f, e^{in\pi x/l}\rangle}{\|e^{-in\pi x/l}\|^2} = \frac{1}{2l}\int_{-l}^{l} f(x)e^{in\pi x/l}dx = \frac{1}{2}(a_n + ib_n), \quad n \in \mathbb{N},$$

so that

$$a_0 + \sum_{n=1}^{N}\left(a_n \cos\frac{n\pi}{l}x + b_n \sin\frac{n\pi}{l}x\right) =$$

$$c_0 + \sum_{n=1}^{N}\left[(c_n + c_{-n})\cos\frac{n\pi}{l}x + i(c_n - c_{-n})\sin\frac{n\pi}{l}x\right]$$

$$= c_0 + \sum_{n=1}^{N}\left(c_n e^{in\pi x/l} + c_{-n}e^{-in\pi x/l}\right) = \sum_{n=-N}^{N} c_n e^{in\pi x/l}.$$

Since $|c_n| \le |a_n| + |b_n| \le 2|c_n|$ for all $n \in \mathbb{Z}$, the two sides of this equation converge or diverge together as $N \to \infty$.

Corollary 3.5 *Any function $f \in \mathcal{L}^2(-l, l)$ can be represented by the Fourier series*

$$\sum_{n=-\infty}^{\infty} c_n e^{in\pi x/l} \equiv c_0 + \sum_{n=1}^{\infty} (c_n e^{in\pi x/l} + c_{-n} e^{-in\pi x/l}), \qquad (3.12)$$

where

$$c_n = \frac{\langle f, e^{in\pi x/l} \rangle}{\|e^{in\pi x/l}\|^2} = \frac{1}{2l} \int_{-l}^{l} f(x) e^{-in\pi x/l} dx, \quad n \in \mathbb{Z}. \qquad (3.13)$$

We saw in Example 1.21 that the set $\{e^{inx} : n \in \mathbb{Z}\}$ is orthogonal in $\mathcal{L}^2(-\pi, \pi)$, hence $\{e^{in\pi x/l} : n \in \mathbb{Z}\}$ is orthogonal in $\mathcal{L}^2(-l, l)$. We can also show that $\{e^{in\pi x/l} : n \in \mathbb{Z}\}$ forms a complete set of eigenfunctions of the SL problem discussed in Example 3.1, where each positive eigenvalue $n^2\pi^2/l^2$ is associated with the pair of eigenfunctions $e^{\pm in\pi x/l}$, and thereby arrive at the representation (3.12) directly. In this representation, the Fourier coefficients c_n are defined by the single formula (3.13) instead of the three formulas for a_0, a_n, and b_n, which is a clear advantage. Using (3.12) and (3.13) to expand the function in Example 3.4, we obtain

$$c_n = \frac{1}{2\pi} \int_{-\pi}^{\pi} f(x) e^{-inx} dx$$

$$= \frac{1}{2\pi} \left[\int_{0}^{\pi} e^{-inx} dx - \int_{-\pi}^{0} e^{-inx} dx \right]$$

$$= \frac{1}{i\pi} \int_{0}^{\pi} \sin nx \, dx$$

$$= \frac{1}{in\pi} (1 - \cos n\pi),$$

$$f(x) = \frac{1}{i\pi} \sum_{n=-\infty}^{\infty} \frac{1}{n} \left[1 - (-1)^n \right] e^{inx}$$

$$= \frac{2}{i\pi} \sum_{n=0}^{\infty} \frac{1}{2n+1} [e^{i(2n+1)x} - e^{-i(2n+1)}]$$

$$= \frac{4}{\pi} \sum_{n=0}^{\infty} \frac{1}{2n+1} \sin(2n+1)x.$$

Exercises

3.1 Verify that the set $\{e^{in\pi x/l} : n \in \mathbb{Z}\}$ is orthogonal in $\mathcal{L}^2(-l, l)$. What is the corresponding orthonormal set?

3.2 Is the series $\sum \dfrac{1}{2n+1} \sin(2n+1)x$ uniformly convergent on $[-\pi, \pi]$? Why?

3.3 Determine the Fourier series for the function f defined on $[-1, 1]$ by

$$f(x) = \begin{cases} 0, & -1 \le x < 0 \\ 1, & 0 \le x \le 1. \end{cases}$$

3.4 Expand the function $f(x) = \pi - |x|$ in a Fourier series on $[-\pi, \pi]$, and prove that the series converges uniformly.

3.5 Expand the function $f(x) = x^2 + x$ in a Fourier series on $[-2, 2]$. Is the convergence uniform?

3.6 If the series $\sum |a_n|$ converges prove that the sum $\sum_{n=1}^{\infty} a_n \cos nx$ is finite for every x in the interval $[-\pi, \pi]$, where it represents a continuous function.

3.7 Prove that the real trigonometric series $\sum (a_n \cos nx + b_n \sin nx)$ converges in $\mathcal{L}^2(-\pi, \pi)$ if, and only if, the numerical series $\sum (a_n^2 + b_n^2)$ converges.

3.2 Pointwise Convergence of Fourier Series

We say that a function $f : \mathbb{R} \to \mathbb{C}$ is *periodic* in p, where p is a positive number, if

$$f(x + p) = f(x) \quad \text{for all } x \in \mathbb{R},$$

and p is then called a *period* of f. When f is periodic in p it is also periodic in any multiple of p, since

$$\begin{aligned} f(x + np) &= f(x + (n-1)p + p) \\ &= f(x + (n-1)p) \\ &= \cdots \\ &= f(x). \end{aligned}$$

In fact the equality $f(x + np) = f(x)$ is true for any $n \in \mathbb{Z}$.

A function f defined on $\mathbb{R}$ which is periodic in p and integrable on $[0, p]$ is clearly integrable over any finite interval and, furthermore, its integral has the same value over all intervals of length p. In other words

$$\int_x^{x+p} f(t)dt = \int_0^p f(t)dt \quad \text{for all } x \in \mathbb{R}. \tag{3.14}$$

This may be proved by using the periodicity of f to show that $\int_0^x f(t)dt = \int_p^{p+x} f(t)dt$ (Exercise 3.8).

When the trigonometric series

$$S_n(x) = a_0 + \sum_{k=1}^{n} (a_k \cos kx + b_k \sin kx)$$

converges on $\mathbb{R}$, its sum

$$S(x) = a_0 + \sum_{k=1}^{\infty} (a_k \cos kx + b_k \sin kx)$$

is clearly a periodic function in 2π, for 2π is a common period of all its terms. If a_k and b_k are the Fourier coefficients of some $\mathcal{L}^2(-\pi, \pi)$ function f,

$$a_0 = \frac{1}{2\pi} \int_{-\pi}^{\pi} f(x)dx,$$

$$a_n = \frac{1}{\pi} \int_{-\pi}^{\pi} f(x) \cos nx \, dx,$$

$$b_n = \frac{1}{\pi} \int_{-\pi}^{\pi} f(x) \sin nx \, dx, \quad n \in \mathbb{N},$$

then we know that S_n converges to f in $\mathcal{L}^2(-\pi, \pi)$,

$$S_n \xrightarrow{\mathcal{L}^2} f.$$

We now wish to investigate the conditions under which S_n converges *pointwise* to f on $[-\pi, \pi]$. More specifically, if f is extended from $[-\pi, \pi]$ to $\mathbb{R}$ as a periodic function by the equation

$$f(x + 2\pi) = f(x) \quad \text{for all } x \in \mathbb{R},$$

when does the equality $f(x) = S(x)$ hold at every $x \in \mathbb{R}$? To answer this question we first need to introduce some definitions.

Definition 3.6

1. A function f defined on a bounded interval I, where $(a, b) \subseteq I \subseteq [a, b]$, is said to be *piecewise continuous* if

    (i) f is continuous on (a, b) except for a finite number of points $\{x_1, x_2, \cdots, x_n\}$,

(ii) the right-hand and left-hand limits

$$\lim_{x \to x_i^+} f(x) = f(x_i^+), \quad \lim_{x \to x_i^-} f(x) = f(x_i^-)$$

exist at every point of discontinuity x_i, and

(iii) $\lim_{x \to a^+} f(x) = f(a^+)$ and $\lim_{x \to b^-} f(x) = f(b^-)$ exist at the end-points of I.

2. f is *piecewise smooth* if f' is piecewise continuous.

3. If the interval I is unbounded, then f is piecewise continuous (smooth) if it is piecewise continuous (smooth) on every bounded subinterval of I.

It is clear from the definition that a piecewise continuous function f on a bounded interval is a locally bounded function whose discontinuities are the result of finite "jumps" in its values. If it is also piecewise smooth then its derivative f' is not defined at a point where f is not continuous, but the right-hand and left-hand limits of f' exist (and may even be equal) at the point of discontinuity. f' is also not defined where it has a jump discontinuity (where f is continuous but its graph has a sharp "corner", such ξ_2 as in Fig. 3.3). Consequently, if f' is piecewise continuous on (a, b), there is a finite sequence of points $\{\xi_i\}$(which include the points of discontinuity of f)

$$a < \xi_1 < \xi_2 < \xi_3 < \cdots < \xi_m < b,$$

where f is not differentiable, but f' is continuous on each open sub-interval (a, ξ_1), (ξ_i, ξ_{i+1}), and (ξ_m, b). Clearly, any piecewise smooth function is piecewise continuous.

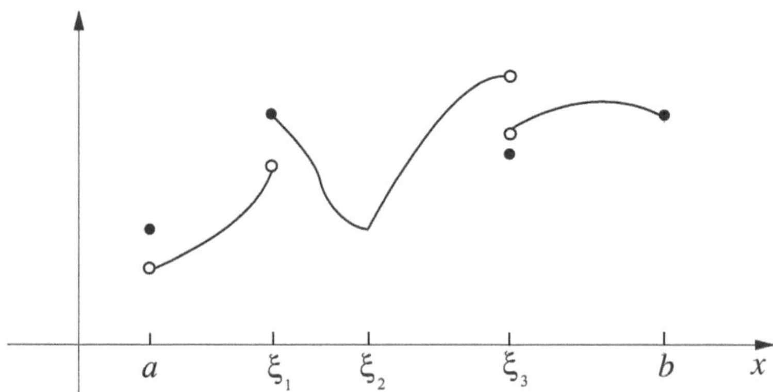

Fig. 3.3 Piecewise smooth function

Note that $f'(x^{\pm})$ are the left-hand and right-hand limits of f' at x, given by

$$f'(x^{-}) = \lim_{x \to x^{-}} f'(x) = \lim_{h \to 0^{+}} \frac{f(x^{-}) - f(x-h)}{h},$$

$$f'(x^{+}) = \lim_{x \to x^{+}} f'(x) = \lim_{h \to 0^{+}} \frac{f(x+h) - f(x^{+})}{h},$$

which are quite different from the left-hand and right-hand derivatives of f at x. The latter are determined by the limits

$$\lim_{h \to 0^{\pm}} \frac{f(x+h) - f(x)}{h},$$

which may not exist at ξ_i even if f is piecewise smooth, such as the step function in Example (3.4). Conversely, not every differentiable function is piecewise smooth (see Exercises 3.10 and 3.11). A continuous function is obviously piecewise continuous, but it may not be piecewise smooth, such as

$$f(x) = \sqrt{x}, \quad 0 \le x \le 1,$$

since $\lim_{x \to 0^{+}} f'(x)$ does not exist.

The next two lemmas will be used to prove Theorem 3.9, which is the main result of this section.

Lemma 3.7 *If f is a piecewise continuous function on $[-\pi, \pi]$, then*

$$\lim_{n \to \infty} \int_{-\pi}^{\pi} f(x) \cos nx dx = \lim_{n \to \infty} \int_{-\pi}^{\pi} f(x) \sin nx dx$$

$$= \lim_{n \to \infty} \int_{-\pi}^{\pi} f(x) e^{\pm inx} dx = 0.$$

Proof Suppose $x_1, \cdots, x_p$ are the points of discontinuity of f on $(-\pi, \pi)$ arranged in increasing order. Since f is continuous and bounded on (x_k, x_{k+1}) for every $0 \le k \le p$, where $x_0 = -\pi$ and $x_{p+1} = \pi$, it is square integrable on all such intervals and

$$\int_{-\pi}^{\pi} |f(x)|^2 dx = \int_{-\pi}^{x_1} |f(x)|^2 dx + \cdots + \int_{x_p}^{\pi} |f(x)|^2 dx.$$

Consequently f belongs to $\mathcal{L}^2(-\pi, \pi)$. Its Fourier coefficients a_n, b_n, and c_n therefore tend to 0 as $n \to \infty$ (see Remark 3.3). $\qquad \square$

Lemma 3.8 *For any real number $\alpha \neq 2m\pi$, $m \in \mathbb{Z}$,*

$$\frac{1}{2} + \cos\alpha + \cos 2\alpha + \cdots + \cos n\alpha = \frac{\sin(n + \frac{1}{2})\alpha}{2\sin\frac{1}{2}\alpha}. \tag{3.15}$$

Proof Using the exponential form of the cosine function, the left-hand side of (3.15) can be written in the form

$$\frac{1}{2} + \sum_{k=1}^{n} \cos k\alpha = \frac{1}{2}\sum_{k=-n}^{n} e^{ik\alpha}$$

$$= \frac{1}{2}e^{-in\alpha}\sum_{k=0}^{2n} e^{ik\alpha}$$

$$= \frac{1}{2}e^{-in\alpha}\sum_{k=0}^{2n} (e^{i\alpha})^k.$$

Since $e^{i\alpha} = 1$ if, and only if, α is an integral multiple of 2π,

$$\frac{1}{2}e^{-in\alpha}\sum_{k=0}^{2n} (e^{i\alpha})^k = \frac{1}{2}e^{-in\alpha}\frac{1 - (e^{i\alpha})^{2n+1}}{1 - e^{i\alpha}}$$

$$= \frac{1}{2}\frac{e^{-in\alpha} - e^{i(n+1)\alpha}}{1 - e^{i\alpha}}, \quad \alpha \neq 0, \pm 2\pi, \cdots. \tag{3.16}$$

Multiplying the numerator and denominator of this last expression by $e^{-i\alpha/2}$, we obtain the right-hand side of (3.15). $\qquad\square$

The expression

$$D_n(\alpha) = \frac{1}{2\pi}\sum_{k=-n}^{n} e^{ik\alpha} = \frac{1}{2\pi} + \frac{1}{\pi}\sum_{k=1}^{n} \cos k\alpha = \frac{1}{2\pi}\frac{e^{i(n+1)\alpha} - e^{-in\alpha}}{e^{i\alpha} - 1},$$

known as the *Dirichlet kernel*, is a continuous function of α which is even and periodic in 2π (Fig. 3.4). Based on Lemma 3.8, the Dirichlet kernel is also represented for all real values of α by

$$D_n(\alpha) = \begin{cases} \dfrac{\sin(n + \frac{1}{2})\alpha}{2\pi\sin\frac{1}{2}\alpha}, & \alpha \neq 0, \pm 2\pi, \cdots \\ \dfrac{2n + 1}{2\pi}, & \alpha = 0, \pm 2\pi, \cdots. \end{cases}$$

Fig. 3.4 Dirichlet kernel D_4

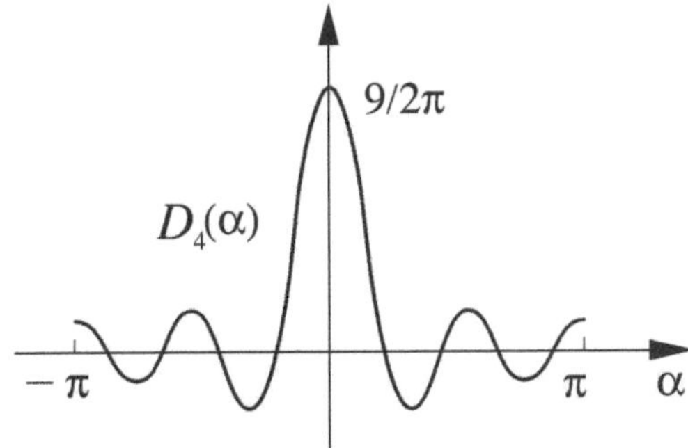

Integrating $D_n(\alpha)$ on $0 \le \alpha \le \pi$, we obtain

$$\int_0^\pi D_n(\alpha)\,d\alpha = \frac{1}{\pi}\int_0^\pi \left(\frac{1}{2} + \sum_{k=1}^n \cos k\alpha\right) d\alpha = \frac{1}{2} \quad \text{for all } n \in \mathbb{N}. \tag{3.17}$$

Now we use Lemmas 3.7 and 3.8 to prove the following pointwise version of the fundamental theorem of Fourier series.

Theorem 3.9 *Let f be a piecewise smooth function on $[-\pi, \pi]$ which is periodic in 2π. If*

$$a_0 = \frac{1}{2\pi}\int_{-\pi}^\pi f(x)\,dx,$$

$$a_n = \frac{1}{\pi}\int_{-\pi}^\pi f(x)\cos nx\,dx,$$

$$b_n = \frac{1}{\pi}\int_{-\pi}^\pi f(x)\sin nx\,dx, \quad n \in \mathbb{N},$$

then the Fourier series

$$S(x) = a_0 + \sum_{n=1}^\infty (a_n \cos nx + b_n \sin nx),$$

converges at every x in $\mathbb{R}$ to $\frac{1}{2}[f(x^+) + f(x^-)]$.

Proof The n-th partial sum of the Fourier series is

$$S_n(x) = a_0 + \sum_{k=1}^n (a_k \cos kx + b_n \sin kx).$$

Substituting the integral representations of the coefficients into this finite sum, and interchanging the order of integration and summation, we obtain

$$S_n(x) = \frac{1}{\pi} \int_{-\pi}^{\pi} f(\xi) \left[\frac{1}{2} + \sum_{k=1}^{n} (\cos k\xi \cos kx + \sin k\xi \sin kx) \right] d\xi$$

$$= \frac{1}{\pi} \int_{-\pi}^{\pi} f(\xi) \left[\frac{1}{2} + \sum_{k=1}^{n} \cos k(\xi - x) \right] d\xi.$$

By the definition of D_n we can therefore write

$$S_n(x) = \int_{-\pi}^{\pi} f(\xi) D_n(\xi - x) d\xi$$

$$= \int_{-\pi-x}^{\pi-x} f(x+t) D_n(t) dt.$$

As a functions of t, both $f(x+t)$ and $D_n(t)$ are periodic in 2π, so we can use the relation (3.14) to write

$$S_n(x) = \int_{-\pi}^{\pi} f(x+t) D_n(t) dt.$$

On the other hand, in view of Eq. (3.17) and the fact that D_n is an even function,

$$\frac{1}{2}[f(x^+) + f(x^-)] = f(x^-) \int_{-\pi}^{0} D_n(t) dt + f(x^+) \int_{0}^{\pi} D_n(t) dt.$$

Hence

$$S_n(x) - \frac{1}{2}[f(x^+) + f(x^-)] = \int_{-\pi}^{0} [f(x+t) - f(x^-)] D_n(t) dt$$

$$+ \int_{0}^{\pi} [f(x+t) - f(x^+)] D_n(t) dt,$$

and we have to show that this sequence tends to 0 as $n \to \infty$.

Define the function

$$g(t) = \begin{cases} \dfrac{f(x+t) - f(x^-)}{e^{it} - 1}, & -\pi < t < 0 \\[2ex] \dfrac{f(x+t) - f(x^+)}{e^{it} - 1}, & 0 < t < \pi. \end{cases}$$

Using (3.16) we can write

$$S_n(x) - \frac{1}{2}[f(x^+) + f(x^-)] = \int_{-\pi}^{\pi} g(t)(e^{it} - 1)D_n(t)\,dt$$

$$= \frac{1}{2\pi} \int_{-\pi}^{\pi} g(t)(e^{i(n+1)t} - e^{-int})\,dt.$$

Since f is piecewise smooth on $[-\pi, \pi]$, the function g is also piecewise smooth on $[-\pi, \pi]$ except possibly at $t = 0$. At $t = 0$ we can use L'Hôpital's rule to obtain

$$\lim_{t \to 0^\pm} g(t) = \lim_{t \to 0^\pm} \frac{f'(x + t)}{i e^{it}} = -i f'(x^\pm).$$

Consequently g is piecewise continuous on $[-\pi, \pi]$, its Fourier coefficients converge to 0 by Lemma 3.7, and therefore

$$\frac{1}{2\pi} \int_{-\pi}^{\pi} g(t) e^{\pm int}\,dt \to 0 \ \text{ as } n \to \infty.$$

$$\square$$

Remark 3.10

1. If f is periodic in 2π and continuous on $[-\pi, \pi]$ then it is continuous on $\mathbb{R}$, and the Fourier series converges pointwise to f on $\mathbb{R}$. Otherwise, where f is discontinuous at a point x, its Fourier series converges to the median of the "jump" at x, which is $\frac{1}{2}[f(x^+) + f(x^-)]$, regardless of how f is defined at that point. This is consistent with the observation that, since the Fourier series S is completely determined by its coefficients a_n and b_n, and since these coefficients are defined by *integrals* involving f, the coefficients would not be sensitive to a change in the value of $f(x)$ at isolated points. If f is defined at its points of discontinuity by the formula

$$f(x) = \frac{1}{2}[f(x^+) + f(x^-)],$$

   then we would have the pointwise equality $S(x) = f(x)$ on the whole of $\mathbb{R}$.

2. Using the exponential form of the Fourier series, we have for any function f which satisfies the hypothesis of the theorem,

$$\lim_{n \to \infty} \sum_{k=-n}^{n} c_k e^{ikx} = \frac{1}{2}[f(x^+) + f(x^-)] \ \text{ for all } x \in \mathbb{R},$$

where

$$c_n = \frac{1}{2\pi} \int_{\pi}^{\pi} f(x)e^{-inx}dx.$$

3. Theorem 3.9 gives sufficient conditions for the pointwise convergence of the Fourier series of f to $\frac{1}{2}[f(x^+) + f(x^-)]$. For an example of a function which is not piecewise smooth, but which can still be represented by a Fourier series, see Exercise 3.26. See [4] and [19] for more details on pointwise convergence of Fourier series.

As would be expected, the interval $[-\pi, \pi]$ in this theorem may be replaced by any other finite interval which is symmetric with respect to the origin.

Corollary 3.11 *Let f be a piecewise smooth function on $[-l, l]$ which is periodic in $2l$. If*

$$a_0 = \frac{1}{2l} \int_{-l}^{l} f(x)dx,$$

$$a_n = \frac{1}{l} \int_{-l}^{l} f(x) \cos \frac{n\pi}{l} x \, dx,$$

$$b_n = \frac{1}{l} \int_{-l}^{l} f(x) \sin \frac{n\pi}{l} x \, dx, \quad n \in \mathbb{N},$$

then the Fourier series

$$a_0 + \sum_{n=1}^{\infty} \left(a_n \cos \frac{n\pi}{l} x + b_n \sin \frac{n\pi}{l} x \right)$$

converges at every x in $\mathbb{R}$ to $\frac{1}{2}[f(x^+) + f(x^-)]$.

Proof Setting $g(x) = f(lx/\pi)$ we see that $g(x + 2\pi) = g(x)$, and that g satisfies the conditions of Theorem 3.9. □

In Example 3.4 the function f is piecewise continuous on $(-\pi, \pi]$, hence its periodic extension to $\mathbb{R}$ satisfies the conditions of Theorem 3.9. Note that f is discontinuous at all integral multiples of π, where it has a jump discontinuity of magnitude $|f(x_n^+) - f(x_n^-)| = 2$. At $x = 0$ we have

$$\frac{1}{2}[f(0^+) + f(0^-)] = \frac{1}{2}[1 - 1] = 0,$$

which agrees with the value of its Fourier series

$$S(x) = \frac{4}{\pi} \sum_{n=0}^{\infty} \frac{1}{2n + 1} \sin(2n + 1)x$$

at $x = 0$. Since $f(0)$ was defined to be 0, the Fourier series converges to f at this point and, by periodicity, at all other even integral multiples of π. The same cannot be said of the point $x = \pi$, where $f(\pi) = 1 \neq S(\pi) = 0$. By periodicity the Fourier series does not converge to f at $x = \pm\pi, \pm3\pi, \cdots$. To obtain convergence at these points (and hence at all points in $\mathbb{R}$), we would have to redefine f at π to be

$$f(\pi) = \frac{1}{2}[f(\pi^+) + f(\pi^-)] = 0.$$

In this same example, since f is continuous at $x = \pi/2$, we have $S(\pi/2) = f(\pi/2) = 1$. Hence

$$1 = \frac{4}{\pi} \sum_{n=0}^{\infty} \frac{1}{2n+1} \sin(2n+1)\frac{\pi}{2}$$

$$= \frac{4}{\pi} \sum_{n=0}^{\infty} \frac{1}{2n+1}(-1)^n,$$

which yields the following series representation of π :

$$\pi = 4\left(1 - \frac{1}{3} + \frac{1}{5} - \frac{1}{7} + \cdots\right). \tag{3.18}$$

Given any real function f, we use the symbol f^+ to denote the positive part of f, that is

$$f^+(x) = \begin{cases} f(x) & \text{if } f(x) \geq 0 \\ 0 & \text{if } f(x) < 0 \end{cases}$$

$$= \frac{1}{2}[f(x) + |f(x)|].$$

Similarly, the negative part of f is

$$f^-(x) = \frac{1}{2}[|f(x)| - f(x)] = \begin{cases} -f(x) & \text{if } f(x) \leq 0 \\ 0 & \text{if } f(x) > 0 \end{cases},$$

and we clearly have $f = f^+ - f^-$.

Example 3.12 The function $f(t) = (2\cos 100\pi t)^+$, shown graphically in Fig. 3.5, describes the current flow, as a function of time t, through an electric semiconductor, also known as a half-wave rectifier. The current amplitude is 2 and its frequency is 50 cycles per second.

To expand f, which is clearly piecewise smooth, in a Fourier series we first have to determine its period. This can be done by noting that the period of $(\cos 100\pi t)^+$

Fig. 3.5 Rectified wave

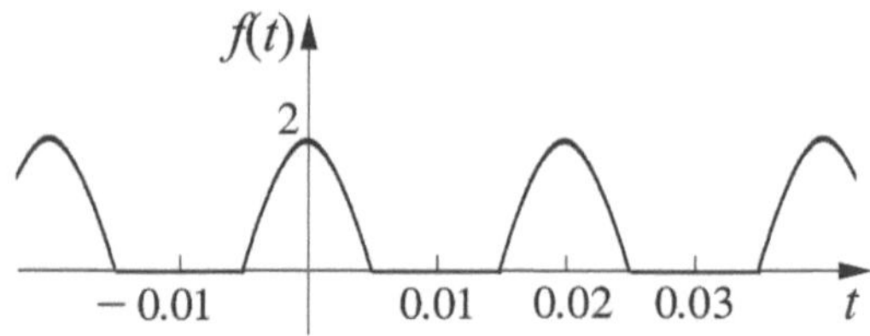

is $2\pi/100\pi = 0.02 = 2l$, from which we conclude that $l = 1/100$. f being even, $b_n = 0$ for all $n \in \mathbb{N}$ and a_n are given by

$$a_0 = \frac{1}{l}\int_0^l f(t)dt = 100\int_0^{1/200} 2\cos 100\pi t\, dt = 2/\pi,$$

$$a_n = \frac{2}{l}\int_0^l f(t)\cos\frac{n\pi}{l}t\, dt$$

$$= 200\int_0^{1/200} 2\cos 100\pi t \cos 100n\pi t\, dt$$

$$= 200\int_0^{1/200} [\cos(n+1)100\pi t + \cos(n-1)100\pi t]dt, \quad n \in \mathbb{N}.$$

When $n = 1$,

$$a_1 = 200\int_0^{1/200} (\cos 200\pi t + 1)dt = 1.$$

For all $n \geq 2$, we have

$$a_n = \frac{2}{\pi}\left[\frac{1}{n+1}\sin(n+1)100\pi t + \frac{1}{n-1}\sin(n-1)100\pi t\right]\Big|_0^{1/200}$$

$$= \frac{2}{\pi}\left[\frac{1}{n+1}\sin(n+1)\frac{\pi}{2} + \frac{1}{n-1}\sin(n-1)\frac{\pi}{2}\right]$$

$$= \frac{2}{\pi}\left(\frac{1}{n+1} - \frac{1}{n-1}\right)\cos n\frac{\pi}{2}$$

$$= -\frac{4}{\pi(n^2-1)}\cos n\frac{\pi}{2}.$$

Thus

$$a_2 = \frac{4}{3\pi}, \quad a_3 = 0, \quad a_4 = -\frac{4}{15\pi}, \quad a_5 = 0, \quad \cdots,$$

and the function f is represented by the Fourier series

$$(2\cos 100\pi t)^+ = \frac{2}{\pi} + \cos 100\pi t + \frac{4}{\pi}\left(\frac{1}{3}\cos 200\pi t - \frac{1}{15}\cos 400\pi t + \cdots\right)$$

$$= \frac{2}{\pi} + \cos 100\pi t - \frac{4}{\pi}\sum_{n=1}^{\infty}\frac{(-1)^n}{4n^2 - 1}\cos 200n\pi t \quad \text{for all } t \in \mathbb{R}.$$

The continuity of f is consistent with the fact that the series converges uniformly (by the M-test). The converse is not true in general, for a convergent series can have a continuous sum without the convergence being uniform. But this cannot happen with a Fourier series, as we now show.

Theorem 3.13 *If f is a continuous function on the interval $[-\pi, \pi]$ such that $f(-\pi) = f(\pi)$ and f' is piecewise continuous on $(-\pi, \pi)$, then the series*

$$\sum_{n=1}^{\infty}\sqrt{|a_n|^2 + |b_n|^2}$$

is convergent, where a_n and b_n are the Fourier coefficients of f defined by

$$a_n = \frac{1}{\pi}\int_{-\pi}^{\pi} f(x)\cos nx \, dx, \quad b_n = \frac{1}{\pi}\int_{-\pi}^{\pi} f(x)\sin nx \, dx.$$

Note that the conditions imposed on f in this theorem are the same as those of Theorem 3.9 with piecewise continuity replaced by continuity on $[-\pi, \pi]$.

Proof Since f' is piecewise continuous on $[-\pi, \pi]$, it belongs to $\mathcal{L}^2(-\pi, \pi)$ and its Fourier coefficients

$$a_0' = \frac{1}{2\pi}\int_{-\pi}^{\pi} f'(x)dx,$$

$$a_n' = \frac{1}{\pi}\int_{-\pi}^{\pi} f'(x)\cos nx \, dx,$$

$$b_n' = \frac{1}{\pi}\int_{-\pi}^{\pi} f'(x)\sin nx \, dx, \quad n \in \mathbb{N},$$

exist. Integrating directly, or by parts, and using the relation $f(-\pi) = f(\pi)$, we obtain

$$a_0' = \frac{1}{2\pi}[f(\pi) - f(-\pi)] = 0,$$

$$a_n' = \frac{1}{\pi}f(x)\cos nx\Big|_{-\pi}^{\pi} + \frac{n}{\pi}\int_{-\pi}^{\pi} f(x)\sin nx \, dx = nb_n,$$

$$b_n' = \frac{1}{\pi} f(x) \sin nx \Big|_{-\pi}^{\pi} - \frac{n}{\pi} \int_{-\pi}^{\pi} f(x) \cos nx \, dx = -na_n.$$

Consequently,

$$
\begin{aligned}
S_N &= \sum_{n=1}^{N} \sqrt{|a_n|^2 + |b_n|^2} \\
&= \sum_{n=1}^{N} \frac{1}{n} \sqrt{|a_n'|^2 + |b_n'|^2} \\
&\le \left[\sum_{n=1}^{N} \frac{1}{n^2} \cdot \sum_{n=1}^{N} \left(|a_n'|^2 + |b_n'|^2 \right) \right]^{1/2},
\end{aligned}
$$

where we use the CBS inequality in the last relation. By Bessel's inequality (1.24) we have

$$\sum_{n=1}^{N} (|a_n'|^2 + |b_n'|^2) \le \int_{-\pi}^{\pi} |f'(x)|^2 \, dx < \infty.$$

Since the series $\sum 1/n^2$ converges, the increasing sequence S_N is bounded above and is therefore convergent. $\qquad\square$

Corollary 3.14 *If f satisfies the conditions of Theorem 3.13, then the convergence of the Fourier series*

$$a_0 + \sum_{n=1}^{\infty} (a_n \cos nx + b_n \sin nx) \tag{3.19}$$

to f on $[-\pi, \pi]$ is uniform and absolute.

Proof The extension of f from $[-\pi, \pi]$ to $\mathbb{R}$ by the periodicity relation

$$f(x + 2\pi) = f(x) \quad \text{for all } x \in \mathbb{R}$$

is a continuous function which satisfies the conditions of Theorem 3.9, hence the Fourier series (3.19) converges to $f(x)$ at every $x \in \mathbb{R}$. The inequality

$$|a_n \cos nx + b_n \sin nx| \le |a_n| + |b_n| \le 2\sqrt{|a_n|^2 + |b_n|^2},$$

and the convergence of $\sum \sqrt{|a_n|^2 + |b_n|^2}$ together imply that the series (3.19) converges uniformly and absolutely by the M-test. $\qquad\square$

Based on Corollaries 1.16 and 3.14, we can now assert that a piecewise smooth function on $[-\pi, \pi]$, which is periodic in 2π, is continuous if, and only if, its Fourier series is uniformly convergent. This result remains valid if the period 2π is replaced by any other period $2l$.

Corollary 3.15 *If f is piecewise smooth on $[-l, l]$ and periodic in $2l$, its Fourier series is uniformly convergent if, and only if, f is continuous.*

Exercises

3.8 Prove Eq. (3.14).

3.9 Determine which of the following functions is piecewise continuous and which is piecewise smooth:

   (a) $f(x) = \sqrt[3]{x}, x \in \mathbb{R}$.
   (b) $f(x) = |\sin x|,\ x \in \mathbb{R}$.
   (c) $f(x) = \sqrt{x}, 0 \le x < 1, f(x+1) = f(x)$ for all $x \ge 0$.
   (d) $f(x) = |x|^{3/2} \sin x, -1 \le x \le 1, f(x+2) = f(x)$ for all $x \in \mathbb{R}$.

3.10 Show that the function $f(x) = |x|$ is piecewise smooth on $\mathbb{R}$, but not differentiable at $x = 0$. Give an example of a function which is piecewise smooth on $\mathbb{R}$ but not differentiable at any integer.

3.11 Show that the function

$$f(x) = \begin{cases} x^2 \sin(1/x), & x \neq 0 \\ 0, & x = 0, \end{cases}$$

is differentiable, but not piecewise smooth, on $\mathbb{R}$.

3.12 Show that the complex Fourier coefficients c_n of an even function are real, and those of an odd function are imaginary.

3.13 Suppose that the functions f and g are piecewise smooth on the bounded interval I. Show that their sum $f + g$ and product $f \cdot g$ are also piecewise smooth on I. What can be said about the quotient f/g?

3.14 Determine the zeros of the Dirichlet kernel D_n and its maximum value.

3.15 Verify that each of the following functions is piecewise smooth and determine its Fourier series:

   (a) $f(x) = x,\ -\pi < x \le \pi,\ f(x + 2\pi) = f(x)$ for all $x \in \mathbb{R}$.
   (b) $f(x) = |x|,\ -1 < x \le 1,\ f(x + 2) = f(x)$ for all $x \in \mathbb{R}$.
   (c) $f(x) = \sin^2 x,\ x \in \mathbb{R}$.
   (d) $f(x) = x^3,\ -l < x \le l,\ f(x + 2l) = f(x)$ for all $x \in \mathbb{R}$.

3.16 In Exercise 3.15 determine where the convergence of the Fourier series is uniform.

3.17 Determine the sum $S(x)$ of the Fourier series at $x = \pm l$ for the function $f(x) = x^3$ in Exercise 3.15(d). Compare with the corresponding result for the function $g(x) = |x|^3$, and explain the difference.

3.18 Use the Fourier series expansion of $f(x) = x$, $-\pi < x \le \pi$, in Exercise 3.15(a) to obtain Eq. (3.18).

3.19 Use the Fourier series expansion of $f(x) = |x|$, $-\pi \le x \le \pi$, $f(x+2\pi) = f(x)$ on $\mathbb{R}$, to obtain a representation of π^2.

3.20 Use the Fourier series expansion of f in Example 3.12 to obtain a representation of π.

3.21 Determine the Fourier series expansion of $f(x) = x^2$ on $[-\pi, \pi]$, then use the result to evaluate each of the following series:

(a) $\displaystyle\sum_{n=1}^{\infty} \frac{1}{n^2}$.

(b) $\displaystyle\sum_{n=1}^{\infty} (-1)^{n+1} \frac{1}{n^2}$.

3.22 Expand the function $f(x) = |\sin x|$ in a Fourier series on $\mathbb{R}$. Verify that the series is uniformly convergent, and use the result to evaluate the series $\displaystyle\sum_{n=1}^{\infty} (4n^2 - 1)^{-1}$ and $\displaystyle\sum_{n=1}^{\infty} (-1)^{n+1} (4n^2 - 1)^{-1}$.

3.23 Show that the function $f(x) = |\sin x|$ has a piecewise smooth derivative. Expand f' in a Fourier series and determine the value of the series at $x = n\pi$ and at $x = \pi/2 + n\pi$, $n \in \mathbb{Z}$. Sketch f and f'.

3.24 Suppose f is a piecewise smooth function on $[0, l]$. The *even periodic extension* of f is the function $\tilde{f}_e$ defined on $\mathbb{R}$ by

$$\tilde{f}_e(x) = \begin{cases} f(x), & 0 \le x \le l \\ f(-x), & -l < x < 0, \end{cases}$$

$$\tilde{f}(x + 2l) = \tilde{f}(x), \quad x \in \mathbb{R};$$

and the *odd periodic extension* of f is

$$\tilde{f}_o(x) = \begin{cases} f(x), & 0 \le x \le l \\ -f(-x), & -l < x < 0, \end{cases}$$

$$\tilde{f}(x + 2l) = \tilde{f}(x), \quad x \in \mathbb{R}.$$

(a)   Show that, if f is continuous on $[0, l]$, then $\tilde{f}_e$ is continuous on $\mathbb{R}$, but that $\tilde{f}_o$ is continuous if, and only if, $f(0) = f(l) = 0$.

(b)   Obtain the Fourier series expansions of $\tilde{f}_e$ and $\tilde{f}_o$.

(c)   Given $f(x) = x$ on $[0, 1]$, determine the Fourier series of $\tilde{f}_e$ and $\tilde{f}_o$.

3.25 For each of the following functions $f : \mathbb{R} \to \mathbb{R}$, determine the value of the Fourier series at $x = 0$ and $x = \pi$, and compare the result with $f(0)$ and $f(\pi)$:

(a) Even periodic extension of $\sin x$ from $[0, \pi]$ to $\mathbb{R}$.
(b) Odd periodic extension of $\cos x$ from $[0, \pi]$ to $\mathbb{R}$.
(c) Odd periodic extension of e^{-x} from $[0, \pi]$ to $\mathbb{R}$.

3.26 Show that the Fourier series of $f(x) = |x|^{1/2}$ converges at every point in $[-\pi, \pi]$. Since f is continuous and lies in $\mathcal{L}^2(-\pi, \pi)$, conclude that its periodic extention to $\mathbb{R}$ is continuous and has a Fourier series which converges pointwise to $f(x)$ at every $x \in \mathbb{R}$, although f is not piecewise smooth.

3.3 Boundary-Value Problems

Fourier series play a crucial role in the construction of solutions to boundary-value and initial-value problems for partial differential equations. A solution of a partial differential equation will naturally be a function of more than one variable. When the equation is linear and homogeneous, an effective method for obtaining its solution is by separation of variables. This is based on the assumption that a solution, say $u(x, y)$, can be expressed as a product of a function of x and a function of y,

$$u(x, y) = v(x)w(y). \tag{3.20}$$

After substituting into the partial differential equation, if we can arrange the terms in the resulting equation so that one side involves only the variable x and the other only y, then each side must be a constant. This gives rise to two ordinary differential equations for the functions v and w which are, presumably, easier to solve than the original equation. The loss of generality involved in the initial assumption (3.20) is compensated for by forming a linear combination of all such solutions $v(x)w(y)$ over the permissible values of the parametric constant. Two well-known examples of boundary-value problems which describe physical phenomena will now be given and solved using separation of variables.

3.3.1 The Heat Equation

Consider the *heat equation* in one space dimension

$$\frac{\partial u}{\partial t} = k \frac{\partial^2 u}{\partial x^2}, \quad 0 < x < l, \ t > 0, \tag{3.21}$$

which governs the heat flow along a thin bar of length l, where $u(x, t)$ is the temperature of the bar at point x and time t, and k is a positive physical constant which depends on the material of the bar. To determine u we need to know the boundary conditions at the ends of the bar and the initial distribution of temperature along the bar. Suppose the boundary conditions on u are simply

$$u(0, t) = u(l, t) = 0, \quad t > 0, \tag{3.22}$$

that is, the bar ends are held at 0 temperature, and the initial temperature distribution along the bar is

$$u(x, 0) = f(x), \quad 0 < x < l, \tag{3.23}$$

for some given function f. We wish to determine $u(x, t)$ for all points (x, t) in the strip $(0, l) \times (0, \infty)$.

Let

$$u(x, t) = v(x)w(t).$$

Substituting into (3.21), we obtain

$$v(x)w'(t) = kv''(x)w(t),$$

which, after dividing by kvw, becomes

$$\frac{v''(x)}{v(x)} = \frac{w'(t)}{kw(t)}. \tag{3.24}$$

Equation (3.24) cannot hold over the strip $(0, l) \times (0, \infty)$ unless each side is a constant, say $-\lambda^2$. The reason for choosing the constant to be $-\lambda^2$ instead of λ will be clarified in Remark 3.16. Thus (3.24) breaks down into two ordinary differential equations,

$$v'' + \lambda^2 v = 0, \tag{3.25}$$

$$w' + \lambda^2 kw = 0. \tag{3.26}$$

The solutions of (3.25) and (3.26), are

$$v(x) = a \cos \lambda x + b \sin \lambda x, \tag{3.27}$$

$$w(t) = ce^{-\lambda^2 kt}, \tag{3.28}$$

where a, b, and c are constants of integration. These solutions depend on the parameter λ, and we indicate this dependence by writing v_λ and w_λ. Thus the solutions we seek have the form

$$u_\lambda(x, t) = (a_\lambda \cos \lambda x + b_\lambda \sin \lambda x)e^{-\lambda^2 kt}, \tag{3.29}$$

where a_λ and b_λ are arbitrary constants which also depend on λ.

The first boundary condition in (3.22) implies

$$u_\lambda(0, t) = a_\lambda e^{-\lambda^2 kt} = 0 \ \text{ for all } t \geq 0,$$

from which we conclude that $a_\lambda = 0$. The second boundary condition gives

$$u_\lambda(l, t) = b_\lambda \sin \lambda l \ e^{-\lambda^2 kt} \ \text{ for all } t \geq 0,$$

hence $b_\lambda \sin \lambda l = 0$. Since we cannot allow b_λ to be 0, otherwise we get the trivial solution $u_\lambda \equiv 0$ which cannot satisfy the initial condition (unless $f \equiv 0$), we conclude that

$$\sin \lambda l = 0,$$

and therefore λl is an integral multiple of π. Thus the parameter λ can only assume the discrete values

$$\lambda_n = \frac{n\pi}{l}, \quad n \in \mathbb{N}.$$

Notice that we have retained only the positive values of n, since the negative values have the effect of changing the sign of $\sin \lambda_n x$, and can therefore be accommodated by the constants $b_{\lambda_n} = b_n$. The case $n = 0$ yields the trivial solution. Thus we arrive at the sequence of solutions to the heat equation, all satisfying the boundary conditions (3.22),

$$u_n(x, t) = b_n \sin \frac{n\pi}{l} x \ e^{-(n\pi/l)^2 kt}, \quad n \in \mathbb{N}.$$

Since Eqs. (3.21) and (3.22) are linear and homogeneous, the formal sum

$$u(x, t) = \sum_{n=1}^{\infty} u_n(x, t) = \sum_{n=1}^{\infty} b_n \sin \frac{n\pi}{l} x \ e^{-(n\pi/l)^2 kt}, \tag{3.30}$$

defines a more general solution of these equations (by the superposition principle).

Now we apply the initial condition (3.23) to the expression (3.30):

$$u(x, 0) = \sum_{n=1}^{\infty} b_n \sin \frac{n\pi}{l} x = f(x).$$

Assuming that the function f is piecewise smooth on $[0, l]$, its odd extension into $[-l, l]$ satisfies the conditions of Corollary 3.11 in the interval $[-l, l]$, and the coefficients b_n must therefore be the Fourier coefficients of f, that is,

$$b_n = \frac{2}{l} \int_0^l f(x) \sin \frac{n\pi}{l} x \, dx.$$

Here we are tacitly assuming that $f(x) = \frac{1}{2}[f(x^+) + f(x^-)]$ for all x, which will be the case if the temperature distribution is continuous. This completely determines $u(x, t)$, represented by the series (3.30), as the solution of the system of Eqs. (3.21), (3.22), and (3.23) in $[0, l] \times [0, \infty)$.

Remark 3.16

1. Although the assumption $u(x, t) = v(x)w(t)$ at the outset places a restriction on the form of the solutions (3.29) that we obtain, the linear combination (3.30) of such solutions restores the generality that was lost, since the sequence $\sin(n\pi x / l)$ spans $\mathcal{L}^2(0, l)$.

2. The reason we chose the constant $v''(x)/v(x) = w'(t)/kw(t)$ to be negative is that a positive constant would change the solutions of Eq. (3.25) from trigonometric to hyperbolic functions, and these cannot be used to expand $f(x)$ (unless λ is imaginary). The square on λ was introduced merely for the sake of convenience, in order to avoid using the square root sign in the representation of the solutions.

In higher space dimensions, the homogeneous heat equation takes the form

$$u_t = k\Delta u, \tag{3.31}$$

where Δ is the *Laplacian operator* in space variables. In $\mathbb{R}^n$ the Laplacian is given in cartesian coordinates by

$$\Delta = \frac{\partial^2}{\partial x_1^2} + \frac{\partial^2}{\partial x_2^2} + \cdots + \frac{\partial^2}{\partial x_n^2}.$$

Example 3.17 The dynamic (i.e. time-dependent) temperature distribution on a rectangular conducting sheet of length a and width b, whose boundary is held at 0 temperature, is described by the following boundary-value problem for the heat equation in $\mathbb{R}^2$:

$$u_t = k(u_{xx} + u_{yy}), \quad (x, y) \in (0, a) \times (0, b), \ t > 0,$$

$$u(0, y, t) = u(a, y, t) = 0, \quad y \in (0, b), \ t > 0,$$

$$u(x, 0, t) = u(x, b, t) = 0, \quad x \in (0, a), \ t > 0,$$

$$u(x, y, 0) = f(x, y), \quad (x, y) \in (0, a) \times (0, b),$$

where we use the more compact notation

$$u_t = \frac{\partial u}{\partial t}, \ u_x = \frac{\partial u}{\partial x}, \ u_{xx} = \frac{\partial^2 u}{\partial x^2}, \ u_{xy} = \frac{\partial^2 u}{\partial y \partial x}, \ u_{yy} = \frac{\partial^2 u}{\partial y^2}, \ \cdots .$$

The assumption that

$$u(x, y, t) = v(x, y)w(t)$$

leads to the equation $v(x, y)w'(t) = kw(t)[v_{xx}(x, y) + v_{yy}(x, y)]$. After dividing by vw and noting that the left-hand side of the resulting equation depends on t and the right-hand side depends on (x, y), we again assume a separation constant $-\lambda^2$ and obtain the pair of equations

$$w' + \lambda^2 kw = 0, \tag{3.32}$$

$$v_{xx} + v_{yy} + \lambda^2 v = 0. \tag{3.33}$$

Equation (3.32), like (3.26), is solved by the exponential function (3.28). We can use separation of variables again to solve (3.33). Substituting $v(x, y) = X(x)Y(y)$ into (3.33) leads to

$$\frac{X''(x)}{X(x)} + \frac{Y''(y)}{Y(y)} + \lambda^2 = 0,$$

where the variables are separable,

$$\frac{X''(x)}{X(x)} = -\frac{Y''(y)}{Y(y)} - \lambda^2 = -\mu^2,$$

and $-\mu^2$ is the second separation constant. The resulting pair of equations have the general solutions

$$X(x) = A \cos \mu x + B \sin \mu x,$$

$$Y(y) = A' \cos \sqrt{\lambda^2 - \mu^2} y + B' \sin \sqrt{\lambda^2 - \mu^2} y.$$

The boundary conditions at $x = 0$ and $y = 0$ imply $A = A' = 0$. From the condition at $x = a$ we obtain $\mu = \mu_n = n\pi/a$ for all positive integers n, and from the condition at $y = b$ we conclude that

$$\sqrt{\lambda^2 - \mu_n^2} = m\pi/b, \quad m \in \mathbb{N},$$

and therefore

$$\lambda = \lambda_{mn} = \sqrt{\frac{n^2}{a^2} + \frac{m^2}{b^2}}\,\pi.$$

Thus we arrive at the double sequence of functions

$$u_{mn}(x, y, t) = b_{mn}e^{-k\lambda_{mn}^2 t} \sin\frac{n\pi}{a}x \, \sin\frac{m\pi}{b}y, \quad m, n \in \mathbb{N},$$

which satisfy the heat equation and the boundary conditions.

Before applying the initial condition we form the formal double sum

$$u(x, y, t) = \sum_{m,n=1}^{\infty} b_{mn}e^{-k\lambda_{mn}^2 t} \sin\frac{n\pi}{a}x \, \sin\frac{m\pi}{b}y, \tag{3.34}$$

which satisfies the boundary conditions on the sides of the rectangular sheet. Now the condition at $t = 0$ implies

$$f(x, y) = \sum_{m,n=1}^{\infty} b_{mn} \sin\frac{n\pi}{a}x \, \sin\frac{m\pi}{b}y. \tag{3.35}$$

Assuming f is square integrable over the rectangle $(0, a) \times (0, b)$, we first extend f as an odd function of x and y into $(-a, a) \times (-b, b)$. Then we multiply Eq. (3.35) by $\sin(n'\pi/a)x$ and integrate over $(-a, a)$ to obtain

$$\sum_{m=1}^{\infty} b_{mn'} \sin\frac{m\pi}{b}y \left\| \sin\frac{n'\pi}{a}x \right\|^2 = 2\int_0^a f(x, y) \sin\frac{n'\pi}{a}x \, dx.$$

Multiplying by $\sin(m'\pi/b)y$ and integrating over $(-b, b)$ yields

$$b_{m'n'} \left\| \sin\frac{n'\pi}{a}x \right\|^2 \left\| \sin\frac{m'\pi}{b}y \right\|^2 = 4\int_0^b\int_0^a f(x, y) \sin\frac{n'\pi}{a}x \, \sin\frac{m'\pi}{b}y \, dxdy.$$

Since

$$\left\| \sin(n'\pi/a)x \right\|^2 = \int_{-a}^{a} \sin^2\left(\frac{n'\pi}{a}x\right) dx = a$$

and

$$\left\| \sin(m'\pi/b)y \right\|^2 = \int_{-b}^{b} \sin^2\left(\frac{m'\pi}{a}y\right) dy = b,$$

the coefficients in (3.34) are given by

$$b_{mn} = \frac{4}{ab} \int_0^a \int_0^b f(x, y) \sin \frac{n\pi}{a} x \, \sin \frac{m\pi}{b} y \, dx dy. \tag{3.36}$$

The expression on the right-hand side of (3.35) is called a double Fourier series of f (see [17]) and the equality holds in $\mathcal{L}^2$ over the rectangle $(0, a) \times (0, b)$. Under appropriate smoothness conditions on the function f, such as the continuity of f and its partial derivatives on $[0, a] \times [0, b]$, (3.34) becomes a pointwise equality in which the Fourier coefficients b_{mn} are defined by (3.36). The series representation (3.34) is then convergent and satisfies the heat equation in the rectangle $(0, a) \times (0, b)$, the conditions on its boundary, and the initial condition at $t = 0$.

It is worth noting that the presence of the exponential function in (3.30) and (3.34) means that the solution of the heat equation decays exponentially to 0 as $t \to \infty$. This is to be expected from a physical point of view, because heat flows from points of higher temperature to points of lower temperature; and since the edges of the sheet are held at (absolute) zero, all heat eventually seeps out. When u does not depend on time, the heat equation becomes Laplace's equation

$$\Delta u = 0.$$

This is a static equation which governs the steady state distribution of potential fields in electrostatics, gravity, fluid flow, and other physical phenomena which can be described by a *potential function*. Some boundary-value problems for the Laplace equation are given in Exercises 3.34 through 3.37. We shall return to this equation in the next chapter.

3.3.2 The Wave Equation

If a thin, flexible, and weightless string, which is stretched horizontally between two fixed points, say $x = 0$ and $x = l$, is given a small vertical displacement and then released from rest, it will vibrate along its length in such a way that the (vertical) displacement at point x and time t, denoted by $u(x, t)$, satisfies the *wave equation* in one space dimension,

$$u_{tt} = c^2 u_{xx}, \quad 0 < x < l, \ t > 0, \tag{3.37}$$

c being a positive constant determined by the material of the string. This equation describes the transverse vibrations of an ideal string, and it differs from the heat equation only in the fact that the time derivative is of second, instead of first, order; but, as we shall soon discover, the solutions are very different. The boundary conditions on u are naturally

$$u(0, t) = u(l, t) = 0, \quad t > 0, \tag{3.38}$$

and the initial conditions are

$$u(x, 0) = f(x), \quad u_t(x, 0) = 0, \quad 0 < x < l, \tag{3.39}$$

$f(x)$ being the initial shape of the string and 0 the initial velocity (as it was released from rest). Here two initial conditions are needed at $t = 0$ because the time derivative in the wave equation is of second order.

Again using separation of variables, with $u(x, t) = v(x)w(t)$, leads to

$$\frac{v''(x)}{v(x)} = \frac{w''(t)}{c^2 w(t)} = -\lambda^2.$$

Hence

$$v''(x) + \lambda^2 v(x) = 0, \quad 0 < x < l,$$
$$w''(t) + c^2 \lambda^2 w(t) = 0, \quad t > 0,$$

whose solutions are

$$v(x) = a \cos \lambda x + b \sin \lambda x,$$
$$w(t) = a' \cos c\lambda t + b' \sin c\lambda t.$$

As in the case of the heat equation, the boundary conditions imply $a = 0$ and $\lambda = n\pi/l$ with $n \in \mathbb{N}$. Thus each function in the sequence

$$u_n(x, t) = \left(a_n \cos \frac{cn\pi}{l} t + b_n \sin \frac{cn\pi}{l} t \right) \sin \frac{n\pi}{l} x, \quad n \in \mathbb{N},$$

satisfies the wave equation and the boundary conditions at $x = 0$ and $x = l$, so (by superposition) the same is true of the (formal) sum

$$u(x, t) = \sum_{n=1}^{\infty} \left(a_n \cos \frac{cn\pi}{l} t + b_n \sin \frac{cn\pi}{l} t \right) \sin \frac{n\pi}{l} x. \tag{3.40}$$

The first initial condition implies

$$\sum_{n=1}^{\infty} a_n \sin \frac{n\pi}{l} x = f(x), \quad 0 < x < l.$$

Assuming f is piecewise smooth, we again invoke Corollary 3.11 to conclude that

$$a_n = \frac{2}{l} \int_0^l f(x) \sin \frac{n\pi}{l} x \, dx. \tag{3.41}$$

The second initial condition gives

$$\sum_{n=1}^{\infty} \frac{cn\pi}{l} b_n \sin \frac{n\pi}{l} x = 0, \quad 0 < x < l,$$

hence $b_n = 0$ for all n.

The solution of the wave Eq. (3.37) under the given boundary and initial conditions (3.38) and (3.39) is therefore

$$u(x, t) = \sum_{n=1}^{\infty} a_n \cos \frac{cn\pi}{l} t \, \sin \frac{n\pi}{l} x, \tag{3.42}$$

where a_n are determined by the Fourier coefficient formula (3.41). Had the string been released with an initial velocity described, for example, by the function

$$u_t(x, 0) = g(x),$$

the coefficients b_n in (3.40) would then be determined by the initial condition

$$u_t(x, 0) = \sum_{n=1}^{\infty} \frac{cn\pi}{l} b_n \sin \frac{n\pi}{l} x = g(x)$$

as

$$b_n = \frac{2}{cn\pi} \int_0^l g(x) \sin \frac{cn\pi}{l} x \, dx.$$

Note how the vibrations of the string, as described by the series (3.42), continue indefinitely as $t \to \infty$. That is because Eq. (3.37) does not take into account any energy loss, such as may be due to air resistance or internal friction in the string. This is in contrast to the solution (3.30) of the heat equation which tends to 0 as $t \to \infty$ as explained above. In Exercise 3.32, where the wave equation includes a friction term due to air resistance, the situation is different.

Exercises

3.27 Use separation of variables to solve the following boundary-value problem for the heat equation:

$$u_t = u_{xx}, \; 0 < x < \pi, \; t > 0,$$

$$u(0, t) = u(\pi, t) = 0, \; t > 0,$$

$$u(x, 0) = \sin^3 x, \; 0 < x < \pi.$$

3.28 If a bar of length l is held at a constant temperature T_0 at one end, and at T_1 at the other, the boundary-value problem becomes

$$u_t = k u_{xx}, \; 0 < x < l, \; t > 0,$$

$$u(0, t) = T_0, \; u(l, t) = T_1, \; t > 0,$$

$$u(x, 0) = f(x), \; 0 < x < l.$$

Solve this system of equations for $u(x, t)$. Hint: Assume that $u(x, t) = v(x, t) + \psi(x)$, where v satisfies the heat equation and the homogeneous boundary-conditions $v(0, t) = v(l, t) = 0$.

3.29 Solve the following boundary-value problem for the wave equation:

$$u_{tt} = u_{xx}, \; 0 < x < l, \; l > 0, \; t > 0,$$

$$u(0, t) = u(l, t) = 0, \; t > 0,$$

$$u(x, 0) = x(l - x), \; u_t(x, 0) = 0, \; 0 < x < l.$$

3.30 Determine the vibrations of the string discussed in Sect. 3.3.2 if it is given an initial velocity $g(x)$ by solving

$$u_{tt} = c^2 u_{xx}, \; 0 < x < l, \; t > 0,$$

$$u(0, t) = u(l, t) = 0, \; t > 0,$$

$$u(x, 0) = f(x), \; u_t(x, 0) = g(x), \; 0 < x < l.$$

Show that the solution can be represented by *d'Alembert's formula*

$$u(x, t) = \frac{1}{2}[\tilde{f}(x + ct) + \tilde{f}(x - ct)] + \frac{1}{2c} \int_{x-ct}^{x+ct} \tilde{g}(\tau) d\tau,$$

where $\tilde{f}$ and $\tilde{g}$ are the odd periodic extensions of f and g, respectively, from $(0, l)$ to $\mathbb{R}$.

3.31 If gravity is taken into account, the vibrations of a string stretched between $x = 0$ and $x = l$ are governed by the equation $u_{tt} = c^2 u_{xx} - g$, where g is the gravitational acceleration constant. Determine these vibrations under the homogeneous boundary conditions $u(0, t) = u(l, t) = 0$, and the homogeneous initial conditions $u(x, 0) = u_t(x, 0) = 0$.

3.32 If the air resistance to a vibrating string is proportional to its velocity, the resulting damped wave equation is

$$u_{tt} + ku_t = c^2 u_{xx}$$

for some constant k. Assuming that the initial shape of the string is $u(x, 0) = 1$, $0 < x < 10$, and that it is released from rest, determine its subsequent shape for all $(x, t) \in (0, 10) \times (0, \infty)$ under homogeneous boundary conditions.

3.33 The vibrations of a rectangular membrane which is fixed along its boundary are described by the system of equations

$$u_{tt} = c^2(u_{xx} + u_{yy}), \quad 0 < x < a, \ 0 < t < b, \ t > 0,$$

$$u(0, y, t) = u(a, y, t) = 0, \quad 0 < y < b, \ t > 0,$$

$$u(x, 0, t) = u(x, b, t) = 0, \quad 0 < x < a, \ t > 0,$$

$$u(x, y, 0) = f(x, y), \ u_t(x, y, 0) = g(x, y), \quad 0 < x < a, \ 0 < y < b.$$

Solve this system of equations in the rectangle $(0, a) \times (0, b)$ for all $t > 0$.

3.34 Determine the solution of Laplace's equation in the rectangle $(0, a) \times (0, b)$ under the conditions $u(0, y) = u(a, y) = 0$ on $0 < y < b$ and $u(x, 0) = f(x)$, $u_y(x, b) = 0$ on $0 < x < a$.

3.35 Solve the boundary-value problem

$$u_{xx} + u_{yy} = 0, \quad 0 < x < 1, \ 0 < y < 1,$$

$$u(0, y) = u_x(1, y) = 0, \quad 0 < y < 1,$$

$$u(x, 0) = 0, \ u(x, 1) = \sin \frac{3\pi}{2} x.$$

3.36 Solve the boundary-value problem

$$u_{xx} + u_{yy} = 0, \quad 0 < x < \pi, \ 0 < y < \pi,$$

$$u_x(0, y) = u_x(\pi, y) = 0, \quad 0 < y < \pi,$$

$$u_y(x, 0) = \cos x, \ u_y(x, \pi) = 0, \quad 0 < x < \pi.$$

3.37 Laplace's equation in polar coordinates (r, θ) is given by

$$u_{rr} + \frac{1}{r} u_r + \frac{1}{r^2} u_{\theta\theta} = 0.$$

(a)   Use separation of variables to show that the solutions of this equation in $\mathbb{R}^2$ are given by

$$u_n(r, \theta) = \begin{cases} a_0 + d_0 \log r, & n = 0 \\ (c_n r^n + d_n r^{-n})(a_n \cos n\theta + b_n \sin n\theta), & n \in \mathbb{N}, \end{cases}$$

where a_n, b_n, and d_n are integration constants. Hint: Use the fact that $u(r, \theta + 2\pi) = u(r, \theta)$.

(b)   Form the general solution $u(r, \theta) = \sum_{n=0}^{\infty} u_n(r, \theta)$ of Laplace's equation, then show that the bounded solution of the equation inside the circle $r = R$ which satisfies $u(R, \theta) = f(\theta)$ is given by

$$u(r, \theta) = A_0 + \sum_{n=1}^{\infty} \left(\frac{r}{R}\right)^n (A_n \cos n\theta + B_n \sin n\theta),$$

for all $r \geq R$, $0 \leq \theta < 2\pi$, where A_n and B_n are the Fourier coefficients of f.

(c)   Determine the (bounded) solution of Laplace's equation in polar coordinates outside the circle $r = R$ under the same boundary condition $u(R, \theta) = f(\theta)$.

Chapter 4
Orthogonal Polynomials

In this chapter we consider three typical examples of singular SL problems whose eigenfunctions are real polynomials. Each set of eigenfunctions is generated by a particular choice of the coefficients in the eigenvalue equation

$$Lu + \lambda u = (pu')' + ru + \lambda \rho u = 0, \tag{4.1}$$

and of the interval $a < x < b$. In all cases the equation

$$\rho p(u'v - uv')\big|_a^b = 0, \tag{4.2}$$

of course, has to be satisfied by any pair of eigenfunctions u and v in order to ensure self-adjointness for L.

If $\{\varphi_n : n \in \mathbb{N}_0\}$ is a complete set of eigenfunctions of some singular SL problem over (a, b), then, being orthogonal and complete in $\mathcal{L}^2_\rho(a, b)$, the sequence (φ_n) may be used to expand any function $f \in \mathcal{L}^2_\rho(a, b)$ by the formula

$$f(x) = \sum_{n=0}^{\infty} \frac{\langle f, \varphi_n \rangle_\rho}{\left\| \varphi_n \right\|_\rho^2} \varphi_n(x) \tag{4.3}$$

in much the same way that the trigonometric functions $\cos nx$ or $\sin nx$ were used to represent a function in $\mathcal{L}^2(0, \pi)$. Thus we arrive at a generalization of the Fourier series which is associated with the choice $p(x) = 1$, $r(x) = 0$, $\rho(x) = 1$, and $(a, b) = (0, \pi)$ in (4.1). For that reason the right-hand side of Eq. (4.3) may be

© The Author(s), under exclusive license to Springer-Verlag London Ltd., part of Springer Nature 2026

M. A. Al-Gwaiz, *Sturm-Liouville Theory and its Applications*, Springer Undergraduate Mathematics Series, https://doi.org/10.1007/978-1-4471-7610-7_4

considered a *generalized Fourier series* of f, and

$$c_n = \frac{\langle f, \varphi_n \rangle_\rho}{\|\varphi_n\|_\rho^2}, \quad n \in \mathbb{N}_0,$$

its *generalized Fourier coefficients*.

A corresponding result to Theorem 3.9 also applies to the generalized Fourier series: If f is piecewise smooth on (a, b), and

$$c_n = \frac{1}{\|\varphi_n\|_\rho^2} \int_a^b f(x)\varphi_n(x)\rho(x)dx,$$

then the series

$$S(x) = \sum_{n=0}^\infty c_n\varphi_n(x)$$

converges at every $x \in (a, b)$ to $\frac{1}{2}[f(x^+) + f(x^-)]$. A general proof of this result is beyond the scope of this treatment. Of course the periodicity property in $\mathbb{R}$, which was peculiar to the trigonometric functions, would not be expected to hold for a general orthogonal basis.

4.1   Legendre Polynomials

The equation

$$(1 - x^2)u'' - 2xu' + \lambda u = 0, \quad -1 < x < 1, \tag{4.4}$$

is called *Legendre's equation*, after the French mathematician A.M. Legendre (1752–1833). It is one of the simplest examples of a singular SL problem, the singularity being due to the fact that $p(x) = 1 - x^2$ vanishes at the end-points $x = \pm 1$. By rewriting Eq. (4.4) in the equivalent form

$$u'' - \frac{2x}{1 - x^2}u' + \frac{\lambda}{1 - x^2}u = 0, \tag{4.5}$$

we see that the coefficients are analytic in the interval $(-1, 1)$, and the solution of the equation can therefore be represented by a power series about the point $x = 0$. Setting

$$u(x) = \sum_{k=0}^\infty c_k x^k, \quad -1 < x < 1, \tag{4.6}$$

and substituting into Legendre's equation, we obtain

$$(1 - x^2) \sum_{k=2}^{\infty} k(k-1)c_k x^{k-2} - 2 \sum_{k=1}^{\infty} k c_k x^k + \lambda \sum_{k=0}^{\infty} c_k x^k = 0$$

$$\sum_{k=0}^{\infty} [(k+2)(k+1)c_{k+2} + (\lambda - k^2 - k)c_k] x^k = 0,$$

from which

$$c_{k+2} = \frac{k(k+1) - \lambda}{(k+1)(k+2)} c_k. \tag{4.7}$$

Equation (4.7) is a recursion formula for the coefficients of the power series (4.6) which expresses the constants c_k for all $k \geq 2$ in terms of c_0 and c_1, which remain arbitrary, and yields two independent power series solutions, one in even powers of x, and the other in odd powers.

If $\lambda = n(n+1)$, where $n \in \mathbb{N}_0$, the recursion relation (4.7) implies

$$0 = c_{n+2} = c_{n+4} = c_{n+6} = \cdots,$$

and it then follows that one of the two solutions is a polynomial. In that case (4.7) takes the form

$$c_{k+2} = \frac{k(k+1) - n(n+1)}{(k+1)(k+2)} c_k = \frac{(k-n)(k+n+1)}{(k+1)(k+2)} c_k. \tag{4.8}$$

Thus, with c_0 and c_1 arbitrary, we have

$$c_2 = -\frac{n(n+1)}{2!} c_0, \qquad c_3 = -\frac{(n-1)(n+2)}{3!} c_1$$
$$c_4 = \frac{n(n-2)(n+1)(n+3)}{4!} c_0, \quad c_5 = \frac{(n-3)(n-1)(n+2)(n+4)}{5!} c_1$$
$$\vdots \qquad\qquad\qquad \vdots$$

and the solution of Legendre's equation takes the form

$$u(x) = c_0 \left[1 - \frac{n(n+1)}{2!} x^2 + \frac{n(n-2)(n+1)(n+3)}{4!} x^4 + \cdots \right] +$$

$$c_1 \left[x - \frac{(n-1)(n+2)}{3!} x^3 + \frac{(n-3)(n-1)(n+2)(n+4)}{5!} x^5 + \cdots \right]$$

$$= c_0 u_0(x) + c_1 u_1(x),$$

where the power series $u_0(x)$ and $u_1(x)$ converge in $(-1, 1)$ and are linearly independent, the first being an even function and the second an odd function.

For each $n \in \mathbb{N}_0$ we obtain a pair of linearly independent solutions,

$$n = 0, \; u_0(x) = 1$$
$$u_1(x) = x + \frac{1}{3}x^3 + \frac{1}{5}x^5 + \cdots,$$
$$n = 1, \; u_0(x) = 1 - x^2 - \frac{1}{3}x^4 + \cdots$$
$$u_1(x) = x,$$
$$n = 2, \; u_0(x) = 1 - 3x^2$$
$$u_1(x) = x - \frac{2}{3}x^3 - \frac{1}{5}x^5 + \cdots,$$
$$n = 3, \; u_0(x) = 1 - 6x^2 + 3x^4 + \cdots$$
$$u_1(x) = x - \frac{5}{3}x^3,$$
$$\vdots \qquad \vdots$$

one of which is a polynomial, and the other an infinite power series which converges in $(-1, 1)$. We are mainly interested in the polynomial solution, which can be written in the form

$$a_n x^n + a_{n-2} x^{n-2} + a_{n-4} x^{n-4} + \cdots. \tag{4.9}$$

This is a polynomial of degree n which is either even or odd, depending on the integer n. By defining the coefficient of the highest power in the polynomial to be

$$a_n = \frac{(2n)!}{2^n (n!)^2} = \frac{1 \cdot 3 \cdot 5 \cdots \cdots (2n-1)}{n!}, \tag{4.10}$$

the resulting expression is called *Legendre's polynomial* of degree n, and is denoted by $P_n(x)$. As a result of this choice, we show in the next section that $P_n(1) = 1$ for all n. The other coefficients in (4.9) are determined in accordance with the recursion relation (4.8):

$$a_{n-2} = -\frac{n(n-1)}{2(2n-1)} a_n$$
$$= -\frac{n(n-1)}{2(2n-1)} \frac{(2n)!}{2^n (n!)^2}$$
$$= -\frac{(2n-2)!}{2^n (n-1)!(n-2)!}$$

$$a_{n-4} = -\frac{(n-2)(n-3)}{4(2n-3)}a_{n-2}$$

$$= \frac{(2n-4)!}{2^n 2!(n-2)!(n-4)!}$$

$$\vdots$$

$$a_{n-2k} = (-1)^k \frac{(2n-2k)!}{2^n k!(n-k)!(n-2k)!}, \quad n \geq 2k,$$

where we arrive at the last equality by induction on k. The last coefficient in P_n is given by

$$a_0 = (-1)^{n/2} \frac{n!}{2^n (n/2)!(n/2)!}$$

if n is even, and

$$a_1 = (-1)^{(n-1)/2} \frac{(n+1)!}{2^n (\frac{n-1}{2})!(\frac{n+1}{2})!}$$

when n is odd. Thus we arrive at the following representation of Legendre's polynomial of degree n,

$$P_n(x) = \frac{1}{2^n} \sum_{k=0}^{[n/2]} (-1)^k \frac{(2n-2k)!}{k!(n-k)!(n-2k)!} x^{n-2k}, \tag{4.11}$$

where $[n/2]$ is the integral part of $n/2$, namely $n/2$ if n is even and $(n-1)/2$ if n is odd. The first six Legendre polynomials are (see Fig. 4.1)

$$
\begin{aligned}
P_0(x) &= 1, & P_1(x) &= x, \\
P_2(x) &= \tfrac{1}{2}(3x^2 - 1), & P_3(x) &= \tfrac{1}{2}(5x^3 - 3x), \\
P_4(x) &= \tfrac{1}{8}(35x^4 - 30x^2 + 3), & P_5(x) &= \tfrac{1}{8}(63x^5 - 70x^3 + 15x).
\end{aligned}
$$

The other solution of Legendre's equation, known as the *Legendre function* Q_n, is an infinite series which converges in the interval $(-1, 1)$ and diverges outside it. For $n = 0$ the Legendre function is given by

$$Q_0(x) = x + \frac{1}{3}x^3 + \frac{1}{5}x^5 + \cdots$$

$$= \frac{1}{2} \log \left(\frac{1+x}{1-x} \right).$$

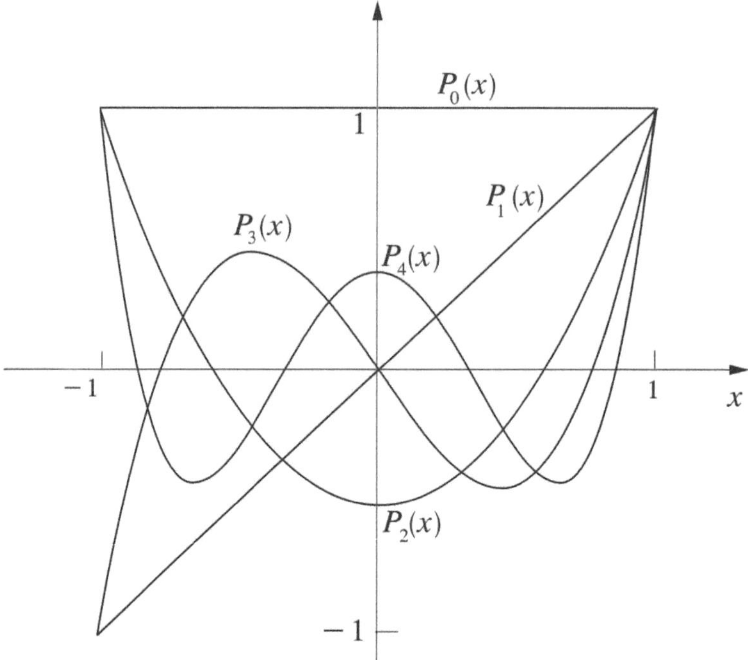

Fig. 4.1 Legendre polynomials P_n

This becomes unbounded as x tends to ± 1 from within the interval $(-1, 1)$. The same is true of $Q_1(x)$ (Exercises 4.3 and 4.4). The other Legendre functions Q_n can also be shown to be singular at the end-points of the interval $(-1, 1)$. The only eigenfunctions of Legendre's equation which are bounded at ± 1 are the Legendre polynomials P_n. Since $p(x) = 1 - x^2 = 0$ at $x = \pm 1$, the condition (4.2) is satisfied.

Thus the differential operator

$$-\frac{d}{dx}\left[(1 - x^2)\frac{d}{dx}\right]$$

in Legendre's equation is self-adjoint. Its eigenvalues $\lambda_n = n(n + 1)$ tend to ∞, and its eigenfunctions in $\mathcal{L}^2(-1, 1)$, namely the Legendre polynomials P_n, are orthogonal and complete in $\mathcal{L}^2(-1, 1)$. We verify the orthogonality of P_n directly in the next section.

Exercises

4.1 Verify that P_n satisfies Legendre's equation when $n = 3$ and $n = 4$.

4.2 Use the Gram-Schmidt method to construct an orthogonal set of polynomials out of the independent set

$$\{1, x, x^2, x^3, x^4, x^5 : -1 \le x \le 1\}.$$

Compare the result with the Legendre polynomials $P_0(x)$, $P_1(x)$, $P_2(x)$, $P_3(x)$, $P_4(x)$, and $P_5(x)$, and show that there is a linear relation between the two sets of polynomials.

4.3 Show that $(-1, 1)$ is the interval of convergence for the power series which defines the Legendre function $Q_n(x)$.

4.4 Prove that

$$Q_1(x) = 1 - \frac{x}{2} \log\left(\frac{1+x}{1-x}\right).$$

4.5 Prove that, for all $n \in \mathbb{N}_0$,

$$P_n(-x) = (-1)^n P_n(x),$$

$$P_{2n+1}(0) = 0,$$

$$P_{2n}(0) = (-1)^n \frac{(2n-1)(2n-3)\cdots(3)(1)}{(2n)(2n-2)\cdots(4)(2)}.$$

4.6 Show that the substitution $x = \cos\theta$, $u(\cos\theta) = y(\theta)$ transforms Legendre's equation to

$$\sin\theta \frac{d^2 y}{d\theta^2} + \cos\theta \frac{dy}{dx} + n(n+1)\sin\theta\, y = 0,$$

where $0 \le \theta \le \pi$. Note the appearance of the weight function $\sin\theta$ in this equation.

4.2   Properties of Legendre Polynomials

For any positive integer n, we can write

$$(x^2 - 1)^n = \sum_{k=0}^{n} (-1)^k \frac{n!}{k!(n-k)!} x^{2n-2k}$$

$$= \sum_{k=0}^{[n/2]} (-1)^k \frac{n!}{k!(n-k)!} x^{2n-2k}$$

$$+ \sum_{k=[n/2]+1}^{n} (-1)^k \frac{n!}{k!(n-k)!} x^{2n-2k}. \tag{4.12}$$

Since

$$k = [n/2] + 1 = \begin{cases} n/2 + 1 & \text{if } n \text{ is even} \\ n/2 + 1/2 & \text{if } n \text{ is odd} \end{cases}$$

implies

$$2n - 2([n/2] + 1) = \begin{cases} n - 2 & \text{if } n \text{ is even} \\ n - 1 & \text{if } n \text{ is odd,} \end{cases}$$

it follows that the powers on x in the second sum on the right-hand side of Eq. (4.12) are all less than n. Taking the n-th derivative of both sides of the equation, therefore, yields

$$\frac{d^n}{dx^n}(x^2 - 1)^n = \frac{d^n}{dx^n} \sum_{k=0}^{[n/2]} (-1)^k \frac{n!}{k!(n-k)!} x^{2n-2k}$$

$$= n! \sum_{k=0}^{[n/2]} (-1)^k \frac{(2n-2k)\cdots(n+1-2k)}{k!(n-k)!} x^{n-2k}$$

$$= n! \sum_{k=0}^{[n/2]} (-1)^k \frac{(2n-2k)!}{k!(n-k)!(n-2k)!} x^{n-2k}.$$

Comparing this with (4.11), we arrive at the *Rodrigues formula* for Legendre polynomials,

$$P_n(x) = \frac{1}{2^n n!} \frac{d^n}{dx^n} (x^2 - 1)^n, \tag{4.13}$$

which provides a convenient method for generating the first few polynomials:

$$P_0(x) = 1,$$

$$P_1(x) = \frac{1}{2} \frac{d}{dx}(x^2 - 1) = x,$$

$$P_2(x) = \frac{1}{8} \frac{d^2}{dx^2}(x^2 - 1)^2 = \frac{1}{2}(3x^2 - 1), \cdots.$$

It can also be used to derive some identities involving the Legendre polynomials and their derivatives, such as

$$P'_{n+1}(x) - P'_{n-1}(x) = (2n+1)P_n(x), \tag{4.14}$$

$$(n+1)P_{n+1}(x) + nP_{n-1}(x) = (2n+1)xP_n(x), \quad n \in \mathbb{N}, \tag{4.15}$$

which will be left as exercises.

We now use the Rodrigues formula to prove the orthogonality of the Legendre polynomials in $\mathcal{L}^2(-1, 1)$. Suppose m and n are any two integers such that $0 \leq m < n$. Using integration by parts, we have

$$\int_{-1}^1 P_n(x)x^m dx = \frac{1}{2^n n!} \int_{-1}^1 x^m \frac{d^n}{dx^n}(x^2 - 1)^n dx$$

$$= \frac{1}{2^n n!} \left[x^m \frac{d^{n-1}}{dx^{n-1}}(x^2 - 1)^n \Big|_{-1}^1 - m \int_{-1}^1 x^{m-1} \frac{d^{n-1}}{dx^{n-1}}(x^2 - 1)^n dx \right]$$

$$= \frac{-m}{2^n n!} x^{m-1} \frac{d^{n-2}}{dx^{n-2}}(x^2 - 1)^n \Big|_{-1}^1 + \frac{m(m-1)}{2^n n!} \int_{-1}^1 x^{m-2} \frac{d^{n-2}}{dx^{n-2}}(x^2 - 1)^n dx.$$

Noting that all derivatives of $(x^2 - 1)^n$ of order less than n vanish at $x = \pm 1$, and repeating this process of integration by parts, we end up in the m-th step with

$$\int_{-1}^1 P_n(x)x^m dx = \frac{(-1)^m m!}{2^n n!} \frac{d^{n-m-1}}{dx^{n-m-1}}(x^2 - 1)^n \Big|_{-1}^1 = 0,$$

because $0 \leq n - m - 1 < n$. Thus we see that P_n is orthogonal to x^m for all $m < n$, which implies that $P_n \perp P_m$ for all $m < n$. By symmetry we therefore conclude that

$$\langle P_n, P_m \rangle = 0 \text{ for all } m \neq n.$$

To evaluate $\|P_n\|$, we use formula (4.13) once more to write

$$\|P_n\|^2 = \int_{-1}^1 P_n^2(x)dx = \frac{1}{2^{2n}(n!)^2} \int_{-1}^1 y^{(n)}(x)y^{(n)}(x)dx, \tag{4.16}$$

where $y(x) = (x^2 - 1)^n$, and then integrate by parts:

$$\int_{-1}^1 y^{(n)}(x)y^{(n)}(x)dx = -\int_{-1}^1 y^{(n-1)}(x)y^{(n+1)}(x)dx$$

$$= \cdots$$

$$= (-1)^n \int_{-1}^{1} y(x) y^{(2n)}(x) dx$$

$$= (-1)^n (2n)! \int_{-1}^{1} y(x) dx. \tag{4.17}$$

$$(-1)^n \int_{-1}^{1} (x^2 - 1)^n dx = \int_{-1}^{1} (1-x)^n (1+x)^n dx$$

$$= \frac{n}{n+1} \int_{-1}^{1} (1-x)^{n-1}(1+x)^{n+1} dx$$

$$= \cdots$$

$$= \frac{n!}{(n+1)\cdots(2n)} \int_{-1}^{1} (1+x)^{2n} dx$$

$$= \frac{(n!)^2}{(2n)!} \frac{(1+x)^{2n+1}}{2n+1} \Big|_{-1}^{1}$$

$$= \frac{(n!)^2 2^{2n+1}}{(2n)!(2n+1)}. \tag{4.18}$$

Combining Eqs. (4.16), (4.17), and (4.18), we get

$$\|P_n\| = \sqrt{\frac{2}{2n+1}}, \quad n \in \mathbb{N}_0,$$

and thereforee the sequence of polynomials

$$\frac{1}{\sqrt{2}} P_0(x), \quad \sqrt{\frac{3}{2}} P_1(x), \quad \sqrt{\frac{5}{2}} P_2(x), \quad \cdots, \quad \sqrt{\frac{2n+1}{2}} P_n(x), \quad \cdots$$

is a complete orthonormal system in $\mathcal{L}^2(-1, 1)$. Note how $\|P_n\|$, unlike $\|\sin nx\|$ and $\|\cos nx\|$, actually depend on n and tend to 0 as $n \to \infty$.

Example 4.1 Since the function

$$f(x) = \begin{cases} 0, & -1 < x < 0 \\ 1, & 0 < x < 1 \end{cases}$$

is square integrable on $(-1, 1)$, it can be represented in $\mathcal{L}^2(-1, 1)$ by the Fourier-Legendre series

$$f(x) = \sum_{n=0}^{\infty} c_n P_n(x),$$

in which the Fourier-Legendre coefficients c_n are determined, according to the expansion formula (4.3), by

$$c_n = \frac{\langle f, P_n \rangle}{\| P_n \|^2} = \frac{2n+1}{2} \int_0^1 f(x) \, P_n(x) dx.$$

The series $\sum c_n P_n$ is also referred to as a Legendre series and c_n its Legendre coefficients. Thus

$$f(x) = \frac{1}{2} P_0(x) + \frac{3}{4} P_1(x) - \frac{7}{16} P_3(x) + \cdots . \tag{4.19}$$

Observe that, in as much as

$$f(x) - \frac{1}{2} P_0(x) = \begin{cases} -1/2, & -1 < x < 0 \\ 1/2, & 0 < x < 1 \end{cases}$$

is an odd function, the Legendre series representation of f does not contain any even Legendre polynomials except P_0, so its value at $x = 0$ is

$$\frac{1}{2} P_0(0) = \frac{1}{2} = \frac{1}{2}[f(0^+) + f(0^-)],$$

as would be expected.

The Legendre polynomials are generated by the function $(1 - 2xt + t^2)^{-1/2}$ in the sense that

$$\frac{1}{\sqrt{1 - 2xt + t^2}} = \sum_{n=0}^{\infty} P_n(x)t^n, \quad |t| < 1, \ |x| \le 1. \tag{4.20}$$

To prove this identity we note that, for every fixed x in $[-1, 1]$, the left-hand side of (4.20) is an analytic function of t in the interval $(-1, 1)$, which we can represent by

$$f(t, x) = \sum_{n=0}^{\infty} a_n(x)t^n, \quad |t| < 1. \tag{4.21}$$

This function satisfies the differential equation

$$(1 - 2xt + t^2)\frac{\partial f}{\partial t} = (x - t)f. \tag{4.22}$$

Substituting the series representation (4.21) into (4.22) leads to the recursion relation

$$(n+1)a_{n+1}(x) + na_{n-1}(x) = (2n+1)xa_n(x), \quad n \in \mathbb{N},$$

which is the same as the relation (4.15) between P_{n+1}, P_{n-1}, and P_n. The first two coefficients in the Taylor series expansion of $(1 - 2xt + t^2)^{1/2}$ are

$$a_0(x) = 1 = P_0(x), \quad a_1(x) = x = P_1(x),$$

and it follows from the recursion relation above that $a_n(x) \equiv P_n(x)$ for all $n \in \mathbb{N}_0$.

Setting $x = 1$ and $x = -1$ in (4.20), we obtain

$$\sum_{n=0}^{\infty} P_n(1)t^n = \frac{1}{\sqrt{1 - 2t + t^2}} = \frac{1}{1-t} = \sum_{n=0}^{\infty} t^n,$$

$$\sum_{n=0}^{\infty} P_n(-1)t^n = \frac{1}{\sqrt{1 + 2t + t^2}} = \frac{1}{1+t} = \sum_{n=0}^{\infty} (-1)^n t^n.$$

Hence

$$P_n(1) = 1, \quad P_n(-1) = (-1)^n \quad \text{for all } n \in \mathbb{N}_0.$$

Exercises

4.7 Use the Rodrigues formula to prove the identity (4.14), then conclude that

$$\int_{\pm 1}^{x} P_n(t)dt = \frac{1}{2n+1}[P_{n+1}(x) - P_{n-1}(x)],$$

$$\int_{-1}^{1} P_n(t)dt = 0 \quad \text{for all } n \in \mathbb{N}.$$

4.8 Prove the identity (4.15).

4.9 Show in detail how Eq. (4.22) follows from Eq. (4.21), and how the two equations imply the recursion relation for a_n.

4.10 Show that if $0 \le r < 1$, then

$$\frac{1}{|1 - re^{i\theta}|} = \sum_{n=0}^{\infty} P_n(\cos\theta)r^n.$$

Hence deduce that, if $\mathbf{x}$ and $\mathbf{y}$ are non-zero vectors in $\mathbb{R}^2$ such that $r = \|\mathbf{x}\| / \|\mathbf{y}\| < 1$ and θ is the angle between them, then

$$\frac{1}{\|\mathbf{y} - \mathbf{x}\|} = \frac{1}{\|\mathbf{y}\|} \sum_{n=0}^{\infty} P_n(\cos\theta) r^n .$$

In particular, if $\|\mathbf{y}\| = 1$, then

$$\frac{1}{\|\mathbf{y} - \mathbf{x}\|} = \sum_{n=0}^{\infty} P_n(\cos\theta) \|\mathbf{x}\|^n .$$

According to a physical interpretation of this result, if two masses are located at $\mathbf{x}$ and $\mathbf{y}$, then the gravitational potential between them may be represented (up to a multiplicative constant) by a power series in $\|\mathbf{x}\|$ with coefficients $P_n(\cos\theta)$.

4.11 Expand each of the following functions in Legendre polynomials:

(a) $f(x) = 1 - x^3$, $-1 \le x \le 1$,
(b) $f(x) = |x|$, $-1 \le x \le 1$.

Compute the first six terms ($n = 0$ up to $n = 5$) in (b).

4.12 Use Eq. (4.15) to prove that

$$n \|P_n\|^2 = (2n - 1) \langle x P_{n-1}, P_n \rangle , \quad n \|P_{n-1}\|^2 = (2n + 1) \langle x P_n, P_{n-1} \rangle ,$$

and hence conclude that $\|P_n\|^2 = 2/(2n + 1)$.

4.13 Expand the function

$$f(x) = \begin{cases} -1, & -1 < x < 0 \\ 1, & 0 < x < 1 \end{cases}$$

in a Legendre series. Use the formulas in Exercise 4.7 to evaluate the coefficients. Compute the value of the series at $x = 0$ and compare it with the average value of $f(0^+)$ and $f(0^-)$.

4.14 Give two Legendre series expansions of the function $f(x) = xe^{-x}$ on $[0, 1]$, one in even degree polynomials, and the other in odd degree. Is the expansion valid at $x = 0$? Repeat the procedure if the function is $f(x) = e^{-x}$ on $[0, 1]$?

4.3 Hermite and Laguerre Polynomials

4.3.1 Hermite Polynomials

As with Legendre's equation, we can generate another sequence of orthogonal polynomials, the *Hermite polynomials,* by solving a suitable SL differential equation of the type (4.1). But, instead, we shall start by defining these polynomials and then derive the equation that they satisfy, though this is not the natural order in which the theory developed; for both Legendre's equation and *Hermite's equation* come up naturally in mathematical physics, as we show later in this chapter.

For each $n \in \mathbb{N}_0$, the function $H_n : \mathbb{R} \to \mathbb{R}$ is defined by

$$H_n(x) = (-1)^n e^{x^2} \frac{d^n}{dx^n} e^{-x^2}. \tag{4.23}$$

This yields a sequence of polynomials

$$
\begin{aligned}
&H_0(x) = 1, &&H_1(x) = 2x, \\
&H_2(x) = 4x^2 - 2, &&H_3(x) = 8x^3 - 12x, \\
&H_4(x) = 16x^4 - 48x^2 + 12, &&H_5(x) = 32x^5 - 160x^3 + 120x, \\
&\quad\vdots &&\quad\vdots
\end{aligned}
$$

known as Hermite polynomials, named after the French mathematician Charles Hermite (1822–1901). Using the formula (4.23), we arrive at the following properties for the set $\{H_n : n \in \mathbb{N}_0\}$.

1. H_n is a polynomial of degree n.

   Proof

$$\frac{d}{dx} e^{-x^2} = -2x e^{-x^2}$$

$$\frac{d^2}{dx^2} e^{-x^2} = (-2x)^2 e^{-x^2} - 2e^{-x^2}$$

$$\frac{d^3}{dx^3} e^{-x^2} = (-2x)^3 e^{-x^2} + 12x e^{-x^2}$$

$$\vdots$$

   By induction, we obtain

$$\frac{d^n}{dx^n} e^{-x^2} = (-2x)^n e^{-x^2} + p(x) e^{-x^2},$$

where p is a polynomial of degree less than n. Therefore

$$
\begin{aligned}
H_n(x) &= (-1)^n e^{x^2}[(-2x)^n + p(x)]e^{-x^2} \\
&= (2x)^n + (-1)^n p(x).
\end{aligned}
\tag{4.24}
$$

$\square$

2. The set $\{H_n : n \in \mathbb{N}_0\}$ is orthogonal in $\mathcal{L}^2_\rho(\mathbb{R})$, where $\rho = e^{-x^2}$.

   Proof Assuming $m < n$,

$$
\begin{aligned}
\langle H_m, H_n \rangle_\rho &= \int_{-\infty}^{\infty} H_m(x) H_n(x) e^{-x^2} dx \\
&= (-1)^n \int_{-\infty}^{\infty} H_m(x) \frac{d^n}{dx^n} e^{-x^2} dx.
\end{aligned}
$$

Integrating by parts n times, and noting that, for any polynomial p, $p(x)e^{-x^2} \to 0$ as $|x| \to \infty$, we obtain

$$
\langle H_m, H_n \rangle_\rho = (-1)^{2n} \int_{-\infty}^{\infty} \left[\frac{d^n}{dx^n} H_m(x) \right] e^{-x^2} dx = 0.
$$

Consequently, H_m is orthogonal to H_n for all $m < n$ and hence, by the symmetry of the inner product, for all $m \neq n$. $\square$

3. $\|H_n\|^2_\rho = 2^n n! \sqrt{\pi}$.

   Proof

$$
\begin{aligned}
\|H_n\|^2_\rho &= \int_{-\infty}^{\infty} H_n^2(x) e^{-x^2} dx \\
&= (-1)^n \int_{-\infty}^{\infty} H_n(x) \frac{d^n}{dx^n} e^{-x^2} dx \\
&= \int_{-\infty}^{\infty} \left[\frac{d^n}{dx^n} H_n(x) \right] e^{-x^2} dx \\
&= \int_{-\infty}^{\infty} 2^n n! e^{-x^2} dx,
\end{aligned}
$$

where the last equality follows from Eq. (4.24). But

$$
\int_{-\infty}^{\infty} e^{-x^2} dx = \sqrt{\pi}
$$

(see Exercise 4.15), and we arrive at the desired equality. $\square$

4. For every $x \in \mathbb{R}$,

$$e^{2xt-t^2} = \sum_{n=0}^{\infty} \frac{1}{n!} H_n(x) t^n. \tag{4.25}$$

In other words, e^{2xt-t^2} is a generating function for the Hermite polynomials.

Proof For every $x \in \mathbb{R}$ the function $f(t,x) = e^{2xt-t^2}$ is analytic in t over $\mathbb{R}$, and is therefore represented by the power series

$$f(t,x) = \sum_{n=0}^{\infty} \frac{1}{n!} \frac{\partial^n f}{\partial t^n}\bigg|_{t=0} t^n.$$

Now we have to show that $\dfrac{\partial^n f}{\partial t^n}\bigg|_{t=0} = H_n(x)$.

$$\begin{aligned}
\frac{\partial^n f}{\partial t^n}\bigg|_{t=0} &= \frac{\partial^n}{\partial t^n} e^{x^2-(x-t)^2}\bigg|_{t=0} \\[2mm]
&= e^{x^2} \frac{\partial^n}{\partial t^n} e^{-(x-t)^2}\bigg|_{t=0} \\[2mm]
&= (-1)^n e^{x^2} \frac{d^n}{dy^n} e^{-y^2}\bigg|_{y=x} \\[2mm]
&= (-1)^n e^{x^2} \frac{d^n}{dx^n} e^{-x^2} = H_n(x).
\end{aligned}$$

$\square$

Theorem 4.2 *H_n satisfies the second order differential equation*

$$u'' - 2xu' + 2nu = 0, \quad x \in \mathbb{R}. \tag{4.26}$$

Proof Differentiating the identity (4.25) with respect to x, we obtain

$$2t e^{2xt-t^2} = \sum_{n=1}^{\infty} \frac{1}{n!} H_n'(x) t^n. \tag{4.27}$$

Using (4.25) again, the left-hand can be written as

$$2t e^{2xt-t^2} = 2\sum_{n=0}^{\infty} \frac{1}{n!} H_n(x) t^{n+1}$$

$$= 2 \sum_{n=0}^{\infty} \frac{n+1}{(n+1)!} H_n(x) t^{n+1}$$

$$= 2 \sum_{n=1}^{\infty} \frac{n}{n!} H_{n-1}(x) t^n. \tag{4.28}$$

By comparing Eqs. (4.27) and (4.28) we arrive at the relation

$$H_n'(x) = 2n H_{n-1}(x), \quad n \in \mathbb{N}. \tag{4.29}$$

Differentiating (4.25) with respect to t gives

$$2(x-t)e^{2xt-t^2} = \sum_{n=1}^{\infty} \frac{1}{(n-1)!} H_n(x) t^{n-1} = \sum_{n=0}^{\infty} \frac{1}{n!} H_{n+1}(x) t^n,$$

from which

$$2x \sum_{n=0}^{\infty} \frac{1}{n!} H_n(x) t^n = 2 \sum_{n=0}^{\infty} \frac{1}{n!} H_n(x) t^{n+1} + \sum_{n=0}^{\infty} \frac{1}{n!} H_{n+1}(x) t^n$$

$$= 2 \sum_{n=0}^{\infty} \frac{n+1}{(n+1)!} H_n(x) t^{n+1} + \sum_{n=0}^{\infty} \frac{1}{n!} H_{n+1}(x) t^n$$

$$= 2 \sum_{n=1}^{\infty} \frac{n}{n!} H_{n-1}(x) t^n + \sum_{n=0}^{\infty} \frac{1}{n!} H_{n+1}(x) t^n.$$

Comparing the coefficients of t^n on the two sides of this equation, we see that

$$2x H_0(x) = H_1(x), \quad n = 0,$$
$$2x H_n(x) = 2n H_{n-1}(x) + H_{n+1}(x), \quad n \in \mathbb{N}. \tag{4.30}$$

In view of Eq. (4.29), this implies

$$2x H_n(x) = H_n'(x) + H_{n+1}(x),$$

and differentiating this last equation, we end up with

$$H_n''(x) = 2x H_n'(x) + 2 H_n(x) - H_{n+1}'(x)$$
$$= 2x H_n'(x) + 2 H_n(x) - (2n+2) H_n(x)$$
$$= 2x H_n'(x) - 2n H_n(x).$$

$\square$

Equation (4.26) is Hermite's equation, and we have shown that one solution of this equation is the Hermite polynomial H_n. As with Legendre's equation, the other solution turns out to be an analytic function which can be represented by a power series in x, but does not lie in $\mathcal{L}_\rho^2$. Multiplication by e^{-x^2} converts Hermite's equation to the standard Sturm-Liouville form

$$(e^{-x^2}u')' + 2ne^{-x^2}u = 0, \quad x \in \mathbb{R}, \tag{4.31}$$

in which $p(x) = e^{-x^2}$, $\lambda = 2n$, $\rho(x) = e^{-x^2}$, and the differential operator

$$\frac{d}{dx}\left(e^{-x^2}\frac{d}{dx}\right)$$

is formally self-adjoint, and Eq. (4.2) is satisfied because $H_n(x)\,e^{-x^2} \to 0$ as $x \to \pm\infty$. We therefore conclude that the sequence of Hermite polynomials $(H_n : n \in \mathbb{N}_0)$ forms a complete orthogonal system in $\mathcal{L}_\rho^2(\mathbb{R})$.

4.3.2 Laguerre Polynomials

The differential equation

$$xu'' + (1-x)u' + nu = 0, \; 0 < x < \infty, \; n \in \mathbb{N}_0, \tag{4.32}$$

is known as *Laguerre's equation*, after the French mathematician *Edmond Laguerre* (1834–1886). Its standard form

$$(xe^{-x}u')' + ne^{-x}u = 0 \tag{4.33}$$

is obtained by multiplication by e^{-x}, the weight function ρ in this case. Note that $p(x) = xe^{-x}$ vanishes at $x = 0$ and decays exponentially as $x \to \infty$, so no boundary condition on the solution are needed. The non-singular solutions of Laguerre's equation are given, up to an arbitrary multiplicative constant, by the *Laguerre polynomials*

$$L_0(x) = 1,$$

$$L_1(x) = 1 - x,$$

$$L_2(x) = 1 - 2x + \frac{1}{2}x^2,$$

$$L_3(x) = 1 - 3x + \frac{3}{2}x^2 - \frac{1}{6}x^3,$$

$$\vdots$$

$$L_n(x) = \frac{e^x}{n!} \frac{d^n}{dx^n} (x^n e^{-x}), \tag{4.34}$$

where L_n has degree n. Since L_n are eigenfunctions of the SL operator in Eq. (4.33), we would expect to have $\langle L_m, L_n \rangle_{e^{-x}} = 0$ for all $m \neq n$, and this follows immediately from the observation that, for all $m < n$,

$$\begin{aligned}
\langle x^m, L_n \rangle_\rho &= \int_0^\infty x^m L_n(x) e^{-x} dx \\
&= \frac{1}{n!} \int_0^\infty x^m \frac{d^n}{dx^n} (x^n e^{-x}) dx \\
&= (-1)^m \frac{m!}{n!} \int_0^\infty \frac{d^{n-m}}{dx^{n-m}} (x^n e^{-x}) dx = 0.
\end{aligned}$$

Furthermore,

$$\langle x^n, L_n \rangle_\rho = (-1)^n \int_0^\infty x^n e^{-x} dx = (-1)^n n!,$$

and since the coefficient of x^n in L_n is $(-1)^n/n!$, it follows that

$$\|L_n\|_\rho^2 = \langle L_n, L_n \rangle_\rho = \frac{(-1)^n}{n!} \langle x^n, L_n \rangle_\rho = 1$$

for all n, and the Laguerre polynomials are in fact orthonormal and complete in $\mathcal{L}_\rho^2(0, \infty)$.

Exercises

4.15 Show that $\int_{-\infty}^\infty e^{-x^2} dx = \sqrt{\pi}$.

    Hint: Square the integral then transform to polar coordinates to evaluate the double integral.

4.16 Show that the powers of x in the polynomial $H_n(x)$ are even if n is even, and odd if n is odd.

4.17 Prove that, for all $n \in \mathbb{N}_0$,

$$H_{2n}(0) = (-1)^n \frac{(2n)!}{n!}, \quad H_{2n+1}(0) = 0.$$

4.18 Use induction to prove

$$H_n(x) = n! \sum_{k=0}^{[n/2]} (-1)^k \frac{(2x)^{n-2k}}{k!(n-2k)!}.$$

4.19 Expand the function $f(x) = x^m$, $m \in \mathbb{N}_0$, in Hermite polynomials.
4.20 Expand the function

$$f(x) = \begin{cases} 0, & x < 0 \\ 1, & x > 0 \end{cases}$$

in terms of Hermite polynomials. Compute the first five terms.
4.21 Prove that

$$L_n(x) = \sum_{k=0}^{n} (-1)^k \frac{1}{n!} \binom{n}{k} x^k,$$

where $\binom{n}{k} = \dfrac{n!}{k!(n-k)!}$ are the binomial coefficients.
4.22 Express $x^3 - x$ as a linear combination of Laguerre polynomials.
4.23 Expand $f(x) = x^m$, where m is a positive integer, in terms of Laguerre polynomials.
4.24 Expand $f(x) = e^{-x/2}$ in a Laguerre series on $[0, \infty)$.
4.25 Verify that L_n satisfies Eq. (4.32).

4.4   Physical Applications

The orthogonal polynomials discussed in this chapter have historically been intimately tied up with the study of potential fields and, more recently, quantum mechanics. Here we look at two typical examples where the Legendre and Hermite polynomials come up. The first is in the description of an electrostatic field generated by a spherical capacitor, as a result of solving Laplace's equation in spherical coordinates. The second is in representing the wave function for a harmonic oscillator.

4.4.1 Laplace's Equation

In $\mathbb{R}^3$ the second order partial differential equation

$$\frac{\partial^2 u}{\partial x^2} + \frac{\partial^2 u}{\partial y^2} + \frac{\partial^2 u}{\partial z^2} = 0, \tag{4.35}$$

is Laplace's equation, after the French mathematician *Pierre de Laplace* (1749–1827). It provides a mathematical model for a number of significant physical phenomena, such as the distribution of an electrostatic or gravitational field in free space, as pointed out in Sect. 3.3. In this context the scalar function $u = u(x, y, z)$ represents the potential at the point (x, y, z). Under the transformation from cartesian coordinates (x, y, z) to spherical coordinates (r, θ, φ), as shown in (Fig. 4.2),

$$x = r \cos \theta \sin \varphi,$$

$$y = r \sin \theta \sin \varphi,$$

$$z = r \cos \varphi, \quad r \geq 0, -\pi \leq \theta < \pi, 0 \leq \varphi \leq \pi.$$

where $r \geq 0, \ -\pi < \theta \leq \pi, 0 \leq \varphi \leq \pi$, Eq. (4.35) takes the form

$$\frac{\partial}{\partial r}\left(r^2 \frac{\partial u}{\partial r}\right) + \frac{1}{\sin^2 \varphi} \frac{\partial^2 u}{\partial \theta^2} + \frac{1}{\sin \varphi} \frac{\partial}{\partial \varphi}\left(\sin \varphi \frac{\partial u}{\partial \varphi}\right) = 0. \tag{4.36}$$

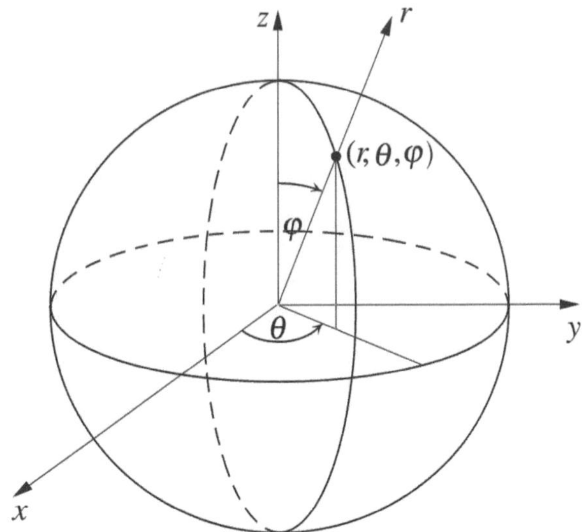

Fig. 4.2 Spherical coordinates

Assuming that the function u is symmetric with respect to the z-axis, it will be independent of θ and Eq. (4.36) becomes

$$\frac{\partial}{\partial r}\left(r^2\frac{\partial u}{\partial r}\right) + \frac{1}{\sin\varphi}\frac{\partial}{\partial\varphi}\left(\sin\varphi\frac{\partial u}{\partial\varphi}\right) = 0. \tag{4.37}$$

Using separation of variables we assume that $u(r, \varphi)$ is a product of a function of r and a function of φ,

$$u(r, \varphi) = v(r)w(\varphi),$$

then substitute into (4.37). After division by vw we obtain

$$\frac{1}{v}\frac{d}{dr}\left(r^2\frac{dv}{dr}\right) = -\frac{1}{w\sin\varphi}\frac{d}{d\varphi}\left(\sin\varphi\frac{dw}{d\varphi}\right), \tag{4.38}$$

where the left-hand side depends only on r, and the right-hand side depends only on φ. Assuming Eq. (4.38) holds over a domain Ω (open connected set) of the $r\varphi$-plane, each side must be a constant, which we denote by λ. Thus we obtain the pair of equations

$$r^2v'' + 2rv' - \lambda v = 0, \tag{4.39}$$

$$\frac{1}{\sin\varphi}(\sin\varphi w')' + \lambda w = 0. \tag{4.40}$$

Setting $\xi = \cos\varphi$ in Eq. (4.40), and noting that

$$\frac{d}{d\varphi} = \frac{d\xi}{d\varphi}\frac{d}{d\xi} = -\sin\varphi\frac{d}{d\xi},$$

$$\frac{1}{\sin\varphi}\frac{d}{d\varphi}\left(\sin\varphi\frac{dw}{d\varphi}\right) = \frac{d}{d\xi}\left[(1 - \xi^2)\frac{dw}{d\xi}\right],$$

leads to Legendre's equation

$$\frac{d}{d\xi}\left[(1 - \xi^2)\frac{dw}{d\xi}\right] + \lambda w = 0.$$

If $\lambda = n(n+1)$, $n \in \mathbb{N}_0$, then (4.40) becomes Legendre's equation (see Exercise 4.6) and its solutions are given by the Legendre polynomials

$$w_n(\varphi) = P_n(\xi) = P_n(\cos\varphi).$$

Equation (4.39), on the other hand, is of the Cauchy-Euler type and the substitution $v(r) = r^\alpha$ yields

$$[\alpha(\alpha - 1) + 2\alpha - n(n + 1)]r^\alpha = (\alpha - n)(\alpha + n + 1)r^\alpha = 0,$$

which implies $\alpha = n$ or $\alpha = -n - 1$. Therefore

$$v_n(r) = a_n r^n + b_n r^{-n-1}$$

for some arbitrary constants a_n and b_n. Thus the sequence

$$u_n(r, \varphi) = v_n(r)w_n(\varphi) = (a_n r^n + b_n r^{-n-1})P_n(\cos \varphi)$$

satisfies Laplace's equation (4.37) for each $n \in \mathbb{N}_0$, and the general solution of the equation is therefore formally given by

$$u(r, \varphi) = \sum_{n=0}^{\infty} u_n(r, \varphi) = \sum_{n=0}^{\infty}(a_n r^n + b_n r^{-n-1})P_n(\cos \varphi). \tag{4.41}$$

Several points are worth noting in connection with this representation. The first is that the Legendre functions $Q_n(\cos \varphi)$ are not taken into account, since these functions become unbounded along the z-axis (where $\cos \varphi = \pm 1$). For the same reason the terms $b_n r^{-n-1}$ have to be dropped (by setting $b_n = 0$) if the domain Ω includes the origin $r = 0$. Once the general solution is formed as a linear combination of all the admissible solutions u_n, the constants a_n and b_n are then determined by the given boundary conditions.

Suppose, for example, that $u(r, \varphi)$ satisfies Laplace's equation inside the ball $0 \leq r < R$ and that $u(R, \varphi) = f(\varphi)$ on the spherical surface $r = R$, where f is a given function in $\mathcal{L}^2(0, \pi)$. Using Eq. (4.41) with $b_n = 0$ for all n,

$$u(r, \varphi) = \sum_{n=0}^{\infty} a_n r^n P_n(\cos \varphi). \tag{4.42}$$

Applying the boundary condition at $r = R$, we obtain

$$u(R, \varphi) = \sum_{n=0}^{\infty} a_n R^n P_n(\cos \varphi) = f(\varphi),$$

from which we conclude that $a_n R^n$ are the Fourier-Legendre coefficients of $f(\varphi)$ when expanded in terms of P_n. Thus

$$a_n R^n = \frac{\langle f, P_n \rangle}{\|P_n\|^2} = \frac{2n + 1}{2} \int_{-1}^{1} f(\varphi(\xi))P_n(\xi)d\xi,$$

and therefore

$$a_n = \frac{2n+1}{2R^n} \int_0^\pi f(\varphi) P_n(\cos \varphi) \sin \varphi \, d\varphi, \quad n \in \mathbb{N}_0. \tag{4.43}$$

Outside the ball, in the region $r > R$, the non-negative powers of r become unbounded and their coefficients must therefore vanish. Hence the solution in that case is given by

$$u(r, \varphi) = \sum_{n=0}^\infty b_n r^{-n-1} P_n(\cos \varphi),$$

and

$$b_n = \frac{2n+1}{2} R^{n+1} \int_0^\pi f(\varphi) P_n(\cos \varphi) \sin \varphi \, d\varphi.$$

Note here that the inner product integral of f and P_n has the weight function $\sin \varphi$, as it should since the Legendre equation now carries this same weight factor (Exercise 4.6).

Example 4.3 If opposite electric charges are placed on two hemispherical conducting sheets, which are insulated from each other, an electric field is generated between them, both inside and outside the spherical surface. The apparatus is called an electric capacitor. Suppose the radius of the spherical capacitor is 1. If the upper hemisphere has potential 1 and the lower hemisphere has potential -1, determine the potential function inside the sphere (Fig. 4.3).

Solution Let u denote the potential function of the electric field at any point in space outside the conducting surfaces. Due to symmetry considerations, the electric charge on each hemisphere (and hence the resulting electric potential) is distributed symmetrically about the z-axis, therefore u depends only on r and φ. Inside the

Fig. 4.3 Spherical capacitor

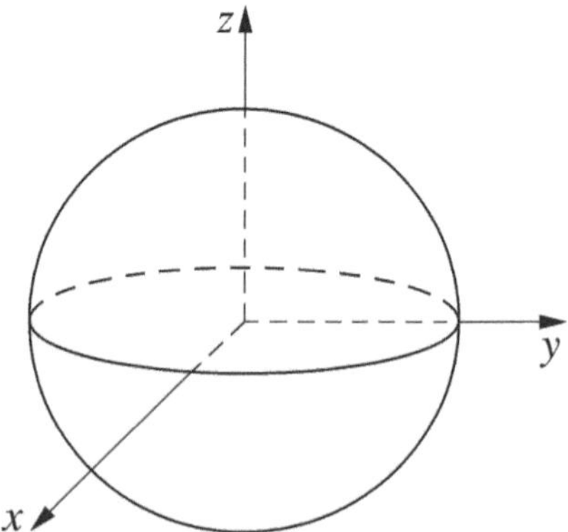

sphere it has the form (4.42), and on the spherical surface

$$u(1, \varphi) = \begin{cases} 1, & 0 \le \varphi < \pi/2 \\ -1, & \pi/2 < \varphi \le \pi. \end{cases}$$

Applying the formula (4.43), with $R = 1$ and $f(\varphi) = u(1, \varphi)$,

$$a_n = \frac{2n+1}{2} \left[\int_0^{\pi/2} P_n(\cos\varphi) \sin\varphi \, d\varphi + \int_{\pi/2}^{\pi} (-1) P_n(\cos\varphi) \sin\varphi \, d\varphi \right]$$

$$= \frac{2n+1}{2} \int_0^1 [P_n(\xi) - P_n(-\xi)] d\xi.$$

Therefore $a_n = 0$ for even values of n, and

$$a_n = \frac{2n+1}{2} \int_0^1 2 P_n(\xi) d\xi = (2n+1) \int_0^1 P_n(\xi) d\xi$$

when n is odd. Thus

$$a_1 = 3 \int_0^1 P_1(\xi) d\xi = \frac{3}{2}$$

$$a_3 = 7 \int_0^1 P_3(\xi) d\xi = -\frac{7}{8}$$

$$a_5 = 11 \int_0^1 P_5(\xi) d\xi = \frac{11}{16}$$

$$\vdots$$

Based on the result of Exercise 4.7, a general formula for a_n, when n is odd, is given by

$$a_n = (2n+1) \int_0^1 P_n(\xi) d\xi$$

$$= P_{n-1}(0) - P_{n+1}(0).$$

The potential inside the capacitor is therefore

$$u(r, \varphi) = \frac{3}{2} r P_1(\cos\varphi) - \frac{7}{8} r^3 P_3(\cos\varphi) + \frac{11}{16} r^5 P_5(\cos\varphi) + \cdots .$$

4.4.2 Harmonic Oscillator

In quantum mechanics the state of a particle which is constrained to move on a straight line (the x-axis) is described by a wave function $\Psi(x, t)$. The motion is said to be *harmonic* if the dependence on time t is given by

$$\Psi(x, t) = \psi(x)e^{-iEt/\hbar},$$

where E is the total energy of the particle and $\hbar$ is a universal constant known as Plank's constant. The location of the particle is then determined by the probability function $\psi(x)$ in the sense that $|\psi(x)|^2 / \|\psi\|^2$ is a measure of its probability density at any point in $\mathbb{R}$. If the particle is located in a potential field $V(x)$, the function ψ satisfies the time-independent *Schrödinger equation*

$$\psi'' + \frac{2m}{\hbar^2}[E - V(x)]\psi = 0.$$

By defining $\lambda = 2mE/\hbar^2$ and $r(x) = -2mV(x)/\hbar^2$, Schrödinger's equation takes the form of a standard Sturm-Liouville eigenvalue equation

$$\psi'' + [\lambda + r(x)]\psi = 0. \tag{4.44}$$

On the interval $-\infty < x < \infty$, the condition that $\psi \to 0$ as $x \to \pm\infty$ defines a singular SL problem.

When $V(x)$ is proportional to x^2, the resulting system is called a *harmonic oscillator*. For the sake of simplicity we take $V(x) = \hbar^2 x^2/2m$, so that Eq. (4.44) becomes

$$\psi'' + (\lambda - x^2)\psi = 0. \tag{4.45}$$

If we set $u(x) = e^{x^2/2}\psi(x)$, then u must satisfy

$$u'' - 2xu' + (\lambda - 1)u = 0,$$

which is Hermite's equation when $\lambda = 2n + 1$, and its solution is then H_n. Consequently, the eigenvalues of Eq. (4.45) are

$$\lambda_n = 2n + 1, \ n \in \mathbb{N}_0,$$

which determine the (admissible) energy levels $E_n = (2n + 1)\hbar^2/2m$ of the particle in the harmonic oscillator, and the corresponding eigenfunctions are given by

$$\psi_n(x) = e^{-x^2/2}H_n(x), \ n \in \mathbb{N}_0,$$

where ψ_n represent the wave function of the particle in energy level E_n. The functions $e^{-x^2/2}H_n(x)$, which belong to $\mathcal{L}^2(\mathbb{R})$, are called *Hermite functions*. The probability that the particle having wave function ψ_n is located in the interval (a, b) is given by

$$\frac{1}{\|H_n\|_\rho^2} \int_a^b H_n^2(x)e^{-x^2}\,dx = \frac{\int_a^b H_n^2(x)e^{-x^2}\,dx}{\int_{-\infty}^\infty H_n^2(x)e^{-x^2}\,dx}.$$

Exercises

4.26 In Example 4.3, determine the surface where $u(r, \varphi) = 0$.

4.27 In Example 4.3, determine the potential function outside the sphere $r = 1$.

4.28 Find the solution $u(r, \varphi)$ of Laplace's equation inside a sphere of radius R if

$$u(R, \varphi) = \begin{cases} 10, & 0 \le \varphi < \pi/2 \\ 0, & \pi/2 < \varphi \le \pi. \end{cases}$$

4.29 Suppose u satisfies Laplace's equation inside the hemisphere $0 \le r < 1, 0 \le \varphi \le \pi/2$. If $u(r, \pi/2) = 0$ on $0 \le r < 1$ and $u(1, \varphi) = 1$ on $0 \le \varphi \le \pi/2$, show that

$$u(r, \varphi) = \sum_{n=0}^\infty (-1)^n \left(\frac{4n+3}{2n+2}\right) \frac{(2n)!}{2^{2n}(n!)^2} r^{2n+1} P_{2n+1}(\cos\varphi)$$

inside the hemisphere.

4.30 Suppose u satisfies Laplace's equation inside the hemisphere $0 \le r < R, 0 \le \varphi \le \pi/2$. If $u_\varphi(r, \pi/2) = 0$ on $0 \le r < R$, and $u(R, \varphi) = f(\varphi)$ on $0 \le \varphi \le \pi/2$, show that

$$u(r, \varphi) = \sum_{n=0}^\infty c_{2n} \left(\frac{r}{R}\right)^{2n} P_{2n}(\cos\varphi),$$

$$c_{2n} = (4n+1) \int_0^{\pi/2} f(\varphi) P_{2n}(\cos\varphi) \sin\varphi\,d\varphi.$$

Chapter 5
Bessel Functions

Bessel functions are another class of eigenfunctions associated with a particular singular SL problem, somewhat different from the polynomials of the last chapter. Here we show how they come up, their special properties, and their applications. But we start by presenting the gamma function, as it plays a role in defining the Bessel functions.

5.1   The Gamma Function

The *gamma function* is defined for all $x > 0$ by the improper integral

$$\Gamma(x) = \int_0^\infty e^{-t} t^{x-1} dt. \tag{5.1}$$

Since this integral converges for all positive values of x, it clearly represents a continuous function on $(0, \infty)$. In fact we can show that Γ is of class C^∞ on $(0, \infty)$ (Exercise 5.1). Integrating by parts, we have

$$\Gamma(x+1) = \int_0^\infty e^{-t} t^x dt = -e^{-t} t^x \big|_0^\infty + x \int_0^\infty e^{-t} t^{x-1} dt$$

$$= x \int_0^\infty e^{-t} t^{x-1} dt,$$

which gives the characteristic recursion relation of the gamma function

$$\Gamma(x+1) = x\Gamma(x), \quad x > 0. \tag{5.2}$$

M. A. Al-Gwaiz, *Sturm-Liouville Theory and its Applications*, Springer Undergraduate Mathematics Series, https://doi.org/10.1007/978-1-4471-7610-7_5

If $x = n$ is a positive integer, then

$$\Gamma(n+1) = n\Gamma(n)$$
$$= n(n-1)\Gamma(n-1)$$
$$\vdots$$
$$= n!\Gamma(1).$$

With

$$\Gamma(1) = \int_0^\infty e^{-t}dt = 1,$$

we have

$$\Gamma(n+1) = n!,$$

which means that the gamma function Γ is an extension of the factorial mapping $n \mapsto (n-1)!$ from $\mathbb{N}$ to $(0, \infty)$.

Equation (5.2) also implies

$$\Gamma(x) = \frac{\Gamma(x+1)}{x},$$

where the right-hand side can be extended to $(-1, 0) \cup (0, \infty)$. Since

$$\lim_{x \to 0} \Gamma(x+1) = \Gamma(1) = 1,$$

the gamma function has a simple pole at $x = 0$. Using the relation (5.2) repeatedly, we see that

$$\Gamma(x) = \frac{\Gamma(x+1)}{x}$$
$$= \frac{\Gamma(x+2)}{x(x+1)}$$
$$\vdots$$
$$= \frac{\Gamma(x+n)}{x(x+1)\cdots(x+n-1)}, \quad n \in \mathbb{N}.$$

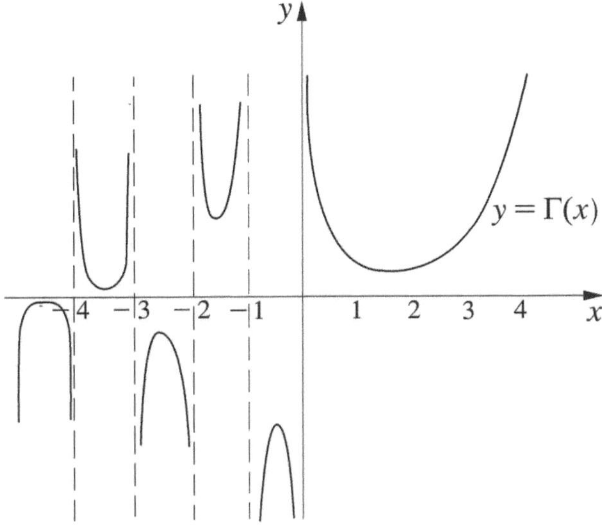

Fig. 5.1 The gamma function

This allows us to extend Γ from $(0, \infty)$ to $\mathbb{R}$, except for the integers $0, -1, -2, \cdots$, where Γ has simple poles (Fig. 5.1). From the inequality

$$\Gamma(x) > \int_1^\infty e^{-t}t^{x-1}dt \geq \int_1^\infty e^{-t}dt = e^{-1}$$

for all $x \geq 1$, and the identity $\Gamma(x+1) = x\Gamma(x)$, we also conclude that $\Gamma(x)$ tends to ∞ as $x \to \infty$.

Exercises

5.1 Prove that the gamma function, as defined in Eq. (5.1), belongs to $C^\infty(0, \infty)$.

5.2 Prove that $\Gamma(1/2) = \sqrt{\pi}$.

5.3 Prove that

$$\Gamma(n + 1/2) = \frac{(2n)!\sqrt{\pi}}{n!2^{2n}}, \ n \in \mathbb{N}_0.$$

5.4 From Exercises 4.15 and 5.2 we have $2\int_0^\infty e^{-x^2}dx = \sqrt{\pi} = \Gamma(1/2)$. Prove that this equality can be generalised to $2\int_0^\infty x^n e^{-x^2}dx = \Gamma((n+1)/2)$ for all $n \in \mathbb{N}_0$.

5.5 The *beta function* is defined by

$$\beta(x, y) = \int_0^1 t^{x-1}(1 - t)^{y-1}dt, \quad x > 0, \, y > 0.$$

Use the transformation $u = t/(1 - t)$ to obtain

$$\beta(x, y) = \int_0^\infty \frac{u^{x-1}}{(1 + u)^{x+y}}du.$$

5.6 The *error function* on $\mathbb{R}$ is defined by the integral

$$\mathrm{erf}(x) = \frac{2}{\sqrt{\pi}} \int_0^x e^{-t^2}dt.$$

Show that it is an odd analytic function on $\mathbb{R}$ and determine its limit as $x \to \pm\infty$. Sketch its graph.

5.2   Bessel Functions of the First Kind

The differential equation

$$x^2y'' + xy' + (x^2 - v^2)y = 0, \tag{5.3}$$

where v is a non-negative parameter, comes up in some situations where separation of variables is used to solve partial differential equations, as we show later in this chapter. It is called *Bessel's equation*, and we shall see that it is another example of an SL eigenvalue equation which generates certain special functions, called *Bessel functions*, in much the same way that the orthogonal polynomials of Chap. 4 were obtained. The main difference is that Bessel functions are not polynomials, and their orthogonality property is somewhat different.

Equation (5.3) has a singular point at $x = 0$, so we cannot expand the solution in a power series about that point. Instead, we use a method due to Georg Frobenius (1849–1917) to construct a solution in terms of real powers (not necessarily integers) of x. The method is based on the premise that every equation of the form

$$y'' + \frac{q(x)}{x}y' + \frac{r(x)}{x^2}y = 0,$$

where the functions q and r are analytic at $x = 0$, has a solution of the form

$$y(x) = x^t \sum_{k=0}^\infty c_k x^k = x^t(c_0 + c_1x + c_2x^2 + \cdots), \tag{5.4}$$

in which t is a real (or complex) number and the constant c_0 is non-zero [12]. Clearly t can always be chosen so that $c_0 \neq 0$. The expression (5.4) becomes a power series when t is a non-negative integer.

Substituting the expression (5.4) into Eq. (5.3), we obtain

$$\sum_{k=0}^{\infty}(k+t)(k+t-1)c_k x^{k+t} + \sum_{k=0}^{\infty}(k+t)c_k x^{k+t}$$

$$+ \sum_{k=0}^{\infty} c_k x^{k+t+2} - v^2 \sum_{k=0}^{\infty} c_k x^{k+t} = 0,$$

or

$$\sum_{k=0}^{\infty}(k+t)^2 c_k x^{k+t} + \sum_{k=0}^{\infty} c_k x^{k+t+2} - v^2 \sum_{k=0}^{\infty} c_k x^{k+t} = 0.$$

Collecting the coefficients of the powers x^t, x^{t+1}, x^{t+2}, $\cdots$, x^{t+j}, we obtain the following equations

$$t^2 c_0 - v^2 c_0 = 0 \tag{5.5}$$

$$(t+1)^2 c_1 - v^2 c_1 = 0 \tag{5.6}$$

$$(t+2)^2 c_2 - v^2 c_2 + c_0 = 0$$

$$\vdots$$

$$(t+j)^2 c_j - v^2 c_j + c_{j-2} = 0. \tag{5.7}$$

From Eq. (5.5) we conclude that $t = \pm v$.

Assuming, to begin with, that $t = v$, Eq. (5.6) becomes

$$(v+1)^2 c_1 - v^2 c_1 = (2v+1)c_1 = 0.$$

Since $2v + 1 \geq 1$, this implies $c_1 = 0$. Now Eq. (5.7) yields

$$[(v+j)^2 - v^2]c_j + c_{j-2} = j(j+2v)c_j + c_{j-2} = 0,$$

and therefore

$$c_j = -\frac{1}{j(j+2v)}c_{j-2}. \tag{5.8}$$

From the equality $c_1 = 0$ it follows that $c_j = 0$ for all odd values of j, and we can assume that $j = 2m$, where m is a positive integer. The recursion relation (5.8) now takes the form

$$c_{2m} = -\frac{1}{2m(2m+2\nu)}c_{2m-2} = -\frac{1}{2^2 m(\nu+m)}c_{2m-2}, \quad m \in \mathbb{N},$$

which allows us to express $c_2, c_4, c_6, \cdots$ in terms of the arbitrary constant c_0:

$$c_2 = -\frac{1}{2^2(\nu+1)}c_0$$

$$c_4 = -\frac{1}{2^2 2(\nu+2)}c_2 = \frac{1}{2^4 2!(\nu+1)(\nu+2)}c_0$$

$$c_6 = -\frac{1}{2^6 3!(\nu+1)(\nu+2)(\nu+3)}c_0$$

$$\vdots$$

$$c_{2m} = \frac{(-1)^m}{2^{2m} m!(\nu+1)(\nu+2)\cdots(\nu+m)}c_0. \tag{5.9}$$

The resulting solution of Bessel's equation is therefore the formal series

$$x^\nu \sum_{m=0}^{\infty} c_{2m} x^{2m}. \tag{5.10}$$

By choosing

$$c_0 = \frac{1}{2^\nu \Gamma(\nu+1)} \tag{5.11}$$

the coefficients (5.9) are given by

$$c_{2m} = \frac{(-1)^m}{2^{\nu+2m} m! \Gamma(\nu+m+1)}.$$

The resulting series

$$x^\nu \sum_{m=0}^{\infty} \frac{(-1)^m}{2^{\nu+2m} m! \Gamma(\nu+m+1)} x^{2m}$$

is called *Bessel's function of the first kind* of order ν, and denoted by $J_\nu(x)$. Thus

$$J_\nu(x) = \left(\frac{x}{2}\right)^\nu \sum_{m=0}^{\infty} \frac{(-1)^m}{m! \Gamma(m+\nu+1)} \left(\frac{x}{2}\right)^{2m}, \tag{5.12}$$

and it is straightforward to verify that the power series

$$\sum_{m=0}^{\infty} \frac{(-1)^m}{2^{2m} m! \, \Gamma(m + \nu + 1)} x^{2m}$$

converges on $\mathbb{R}$ by the ratio test. With $\nu \geq 0$ the power x^ν is well defined when x is positive, hence the Bessel function $J_\nu(x)$ is well defined by (5.12) on $(0, \infty)$. Since

$$\lim_{x \to 0^+} J_\nu(x) = \begin{cases} 1, & \nu = 0 \\ 0, & \nu > 0, \end{cases}$$

the function J_ν may be extended as a continuous function to $[0, \infty)$ by the definition $J_\nu(0) = \lim_{x \to 0^+} J_\nu(x)$ for all $\nu \geq 0$.

Now if we set $t = -\nu < 0$ in (5.4), that is, if we change the sign of ν in (5.12), then

$$J_{-\nu}(x) = \left(\frac{x}{2}\right)^{-\nu} \sum_{m=0}^{\infty} \frac{(-1)^m}{m! \, \Gamma(m - \nu + 1)} \left(\frac{x}{2}\right)^{2m}, \quad x > 0, \tag{5.13}$$

remains a solution of Bessel's equation, since the equation is invariant under such a change of sign. But it will not necessarily be bounded at $x = 0$, as we show in the next theorem.

Theorem 5.1 *The Bessel functions J_ν and $J_{-\nu}$ are linearly independent if, and only if, ν is not an integer.*

Proof If $\nu = n \in \mathbb{N}_0$, then

$$J_{-n}(x) = \left(\frac{x}{2}\right)^{-n} \sum_{m=0}^{\infty} \frac{(-1)^m}{m! \, \Gamma(m - n + 1)} \left(\frac{x}{2}\right)^{2m}.$$

But since $1/\Gamma(m - n + 1) = 0$ for all $m - n + 1 \leq 0$, the terms in which $m = 0, 1, \cdots, n - 1$ all vanish, and we end up with

$$J_{-n}(x) = \left(\frac{x}{2}\right)^{-n} \sum_{m=n}^{\infty} \frac{(-1)^m}{m! \, \Gamma(m - n + 1)} \left(\frac{x}{2}\right)^{2m}$$

$$= \left(\frac{x}{2}\right)^{-n} \sum_{m=0}^{\infty} \frac{(-1)^{m+n}}{(m + n)! \, \Gamma(m + 1)} \left(\frac{x}{2}\right)^{2m+2n}$$

$$= (-1)^n \left(\frac{x}{2}\right)^{n} \sum_{m=0}^{\infty} \frac{(-1)^m}{m! \, \Gamma(m + n + 1)} \left(\frac{x}{2}\right)^{2m}$$

$$= (-1)^n J_n(x).$$

Now suppose $\nu > 0$, $\nu \notin \mathbb{N}_0$, and let

$$a J_\nu(x) + b J_{-\nu}(x) = 0. \tag{5.14}$$

Taking the limit as $t \to 0$ in this equation, we have $\lim_{x\to 0+} J_\nu(x) = 0$ whereas $\lim_{x\to 0+} |J_{-\nu}| = \infty$ because the first term in the series (5.13), which is

$$\frac{1}{\Gamma(1-\nu)} \left(\frac{x}{2}\right)^{-\nu},$$

dominates all the other terms and tends to $\pm\infty$. Thus the equality (5.14) cannot hold on $(0, \infty)$ unless $b = 0$, in which case $a = 0$ as well. $\square$

Based on this theorem we therefore conclude that, when ν is not an integer, the general solution of Bessel's equation on $(0, \infty)$ is given by $y(x) = c_1 J_\nu(x) + c_2 J_{-\nu}(x)$. The general solution when ν is an integer will have to await the definition of Bessel's function of the second kind in Sect. 5.3.

In the following example we prove the first of several identities involving the Bessel functions.

Example 5.2

$$\frac{d}{dx}[x^{-\nu} J_\nu(x)] = -x^{-\nu} J_{\nu+1}(x), \quad x > 0, \nu \geq 0. \tag{5.15}$$

Proof The function $x^{-\nu} J_\nu(x)$, being a power series, can be differentiated term by term:

$$\frac{d}{dx}[x^{-\nu} J_\nu(x)] = \frac{d}{dx} \sum_{m=0}^{\infty} \frac{(-1)^m}{2^{2m+\nu} m! \Gamma(m+\nu+1)} x^{2m}$$

$$= \sum_{m=1}^{\infty} \frac{(-1)^m 2m}{2^{2m+\nu} m! \Gamma(m+\nu+1)} x^{2m-1}$$

$$= -x^{-\nu} \sum_{m=0}^{\infty} \frac{(-1)^m}{2^{2m+\nu+1} m! \Gamma(m+\nu+2)} x^{2m+\nu+1}$$

$$= -x^{-\nu} J_{\nu+1}.$$

$\square$

A corresponding identity to (5.15),

$$\frac{d}{dx}\left[x^\nu J_\nu(x)\right] = x^\nu J_{\nu-1}(x),$$

can similarly be proved (see Exercise 5.11).

Bessel's functions of integral order, given by

$$J_n(x) = \sum_{m=0}^{\infty} \frac{(-1)^m}{m!(m+n)!} \left(\frac{x}{2}\right)^{2m+n}, \quad n \in \mathbb{N}_0, \tag{5.16}$$

are analytic in $(0, \infty)$, and have analytic extensions to $\mathbb{R}$ as even or odd functions, depending on whether n is even or odd. We have already established, in Example 2.13, that J_0 has an infinite set of isolated zeros in $(0, \infty)$ which accumulate at ∞. We can arrange these in an increasing sequence

$$\xi_{01} < \xi_{02} < \xi_{03} < \cdots$$

such that $\xi_{0k} \to \infty$ as $k \to \infty$. Using mathematical induction, we can show that the same is also true of J_n for any positive integer n: Suppose that the zeros of J_m, for any positive integer m, is an increasing sequence $(\xi_{mk} : k \in \mathbb{N})$ in $(0, \infty)$ which tends to ∞. Since the function $x^{-m} J_m(x)$ vanishes at any pair of consecutive zeros, say ξ_{mk} and $\xi_{m\,k+1}$, it follows from Rolle's theorem that there is at least one point between ξ_{mk} and $\xi_{m\,k+1}$ where the derivative of $x^{-m} J_m(x)$ vanishes. In view of the identity (5.15), the function

$$J_{m+1}(x) = -x^m \frac{d}{dx}[x^{-m} J_m(x)]$$

has at least one zero between ξ_{mk} and $\xi_{m\,k+1}$. Thus we have proved

Theorem 5.3 *For any $n \in \mathbb{N}_0$, the equation $J_n(x) = 0$ has an infinite number of positive roots, which form an increasing sequence*

$$\xi_{n1} < \xi_{n2} < \xi_{n3} < \cdots$$

such that $\xi_{nk} \to \infty$ as $k \to \infty$.

The first two Bessel functions of integral order are

$$J_0(x) = \sum_{m=0}^{\infty} \frac{(-1)^m}{(m!)^2} \left(\frac{x}{2}\right)^{2m}$$

$$= 1 - \frac{x^2}{2^2(1!)^2} + \frac{x^4}{2^4(2!)^2} - \frac{x^6}{2^6(3!)^2} + \cdots,$$

$$J_1(x) = \sum_{m=0}^{\infty} \frac{(-1)^m}{m!(m+1)!} \left(\frac{x}{2}\right)^{2m+1}$$

$$= \frac{x}{2} - \frac{x^3}{2^3 1!2!} + \frac{x^5}{2^5 2!3!} - \frac{x^7}{2^7 3!4!} + \cdots.$$

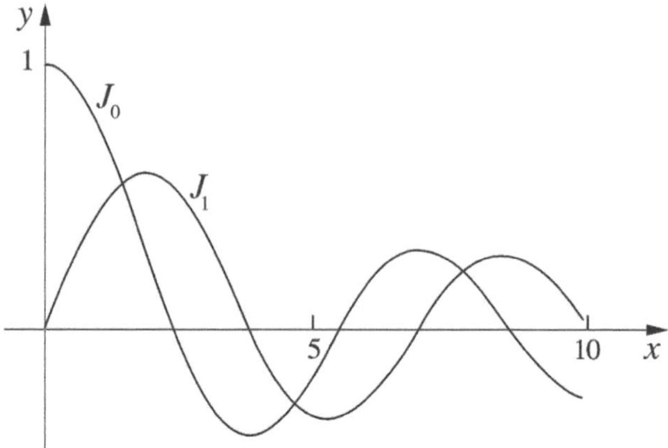

Fig. 5.2 Bessel functions J_0 and J_1

The similarities between these two expansions on the one hand, and those of the cosine and sine functions on the other, are quite striking. The graphs of $J_0(x)$ and $J_1(x)$ in Fig. 5.2 also exhibit many of the properties of $\cos x$ and $\sin x$, respectively, such as the behaviour near $x = 0$ and the interlacing of their zeros. When we set $\nu = 0$ in the identity (5.15) we obtain the relation $J_0'(x) = -J_1(x)$, which corresponds to the familiar relation $(\cos)'x = -\sin x$. But, unlike the situation with trigonometric functions, the distribution of the zeros of J_n is not uniform in general (see, however, Exercise 5.8) and the amplitude of the function decreases with increasing x (Exercise 5.30).

In the next example we prove another important relation between J_0 and J_1, one which we shall have occasion to resort to later in this chapter.

Example 5.4

$$\int_0^x t J_0(t)dt = x J_1(x) \ \text{ for all } x > 0.$$

Proof

$$\int_0^x t J_0(t)dt = \int_0^x \sum_{m=0}^{\infty} \frac{(-1)^m}{(m!)^2 2^{2m}} t^{2m+1} dt$$

$$= \sum_{m=0}^{\infty} \frac{(-1)^m x^{2m+2}}{(m!)^2 (2m+2) 2^{2m}}$$

$$= x \sum_{m=0}^{\infty} \frac{(-1)^m}{m!(m+1)!} \left(\frac{x}{2}\right)^{2m+1} = x J_1(x).$$

$\square$

Exercises

5.7 Verify that the power series which represents $x^{-\nu} J_\nu(x)$ converges on $\mathbb{R}$ for every $\nu \geq 0$.

5.8 Prove that

$$J_{1/2}(x) = \sqrt{\frac{2}{\pi x}} \sin x, \quad J_{-1/2}(x) = \sqrt{\frac{2}{\pi x}} \cos x,$$

and sketch these two functions.

5.9 Prove that $x J_\nu'(x) = \nu J_\nu(x) - x J_{\nu+1}(x)$, and hence the identity of Example 5.2.

5.10 Use Exercises 5.8 and 5.9 to prove that

$$J_{3/2}(x) = \sqrt{\frac{2}{\pi x}} \left(\frac{\sin x}{x} - \cos x \right).$$

5.11 Prove the identity $[x^\nu J_\nu(x)]' = x^\nu J_{\nu-1}(x)$, and hence conclude that

$$J_{-3/2}(x) = -\sqrt{\frac{2}{\pi x}} \left(\frac{\cos x}{x} + \sin x \right).$$

5.12 Use the identities of Example 5.2 and Exercise 5.11 to establish

$$J_\nu'(x) = \frac{1}{2} [J_{\nu-1}(x) - J_{\nu+1}(x)].$$

5.13 Prove that

$$J_{\nu+1}(x) + J_{\nu-1}(x) = \frac{2\nu}{x} J_\nu(x).$$

5.14 Derive the following relations:

(a) $\int_0^x t^2 J_1(t)\,dt = 2x J_1(x) - x^2 J_0(x)$.
(b) $\int_0^x J_3(t)\,dt = 1 - J_2(x) - 2J_1(x)/x$.

5.15 Use the identities $J_0'(x) = -J_1(x)$ and $[x J_1(x)]' = x J_0(x)$ to prove that, for all $n = 2, 3, 4, \cdots$,

$$\int_0^x t^n J_0(t)\,dt = x^n J_1(x) + (n-1)x^{n-1} J_0(x) - (n-1)^2 \int_0^x t^{n-2} J_0(t)\,dt.$$

5.16 Verify that the Wronskian $W(x) = W(J_\nu, J_{-\nu})$, where $\nu \notin \mathbb{N}_0$, satisfies the equation $x\,W' + W = 0$, and thereby prove that

$$W(x) = -\frac{2}{\Gamma(\nu)\Gamma(1-\nu)x}.$$

5.3 Bessel Functions of the Second Kind

Recalling Theorem 5.1, it is natural to ask what the general solution of Bessel's equation would look like when ν is an integer n. There are several ways we can define a second solution to Bessel's equation which is independent of J_n. The more common approach is to define *Bessel's function of the second kind* of order ν by

$$Y_\nu(x) = \begin{cases} \dfrac{1}{\sin \nu\pi}[J_\nu(x)\cos \nu\pi - J_{-\nu}(x)], & \nu \neq 0, 1, 2, \cdots \\ \lim_{\nu \to n} Y_\nu, & n = 0, 1, 2, \cdots. \end{cases}$$

In this connection, observe the following points:

(i) For non-integer values of ν, Y_ν is a linear combination of J_ν and $J_{-\nu}$. Because $J_{-\nu}$ is linearly independent of J_ν, so is Y_ν.

(ii) When $\nu = n$, the above definition gives the indeterminate form $0/0$. By L'Hôpital's rule, using the differentiability properties of power series,

$$\begin{aligned}
Y_n(x) &= \frac{1}{\pi}\left[\left.\frac{\partial J_\nu(x)}{\partial \nu}\right|_{\nu=n} - (-1)^n \left.\frac{\partial J_{-\nu}(x)}{\partial \nu}\right|_{\nu=n}\right] \\
&= \frac{2}{\pi}\left(\log\frac{x}{2} + \gamma\right)J_n(x) - \frac{1}{\pi}\left(\frac{x}{2}\right)^n \sum_{m=0}^{\infty}\frac{(-1)^m(h_m + h_{n+m})}{m!(n+m)!}\left(\frac{x}{2}\right)^{2m} \\
&\quad - \frac{1}{\pi}\left(\frac{x}{2}\right)^{-n}\sum_{m=0}^{n-1}\frac{(n-m-1)!}{m!}\left(\frac{x}{2}\right)^{2m}, \quad x > 0,
\end{aligned}$$

where

$$h_0 = 0, \ h_m = 1 + \frac{1}{2} + \frac{1}{3} + \cdots + \frac{1}{m},$$

and

$$\gamma = \lim_{m\to\infty}(h_m - \log m) = 0.577215\cdots$$

is called Euler's constant. Note that the last sum in the expression for Y_n vanishes when $n = 0$, and that the presence of the term $\log x\, J_n(x)$ implies Y_n is linearly independent of J_n. That the passage to the limit as $\nu \to n$ preserves Y_n as a solution of Bessel's equation is due to the continuity of Bessel's equation and Y_ν with respect to ν. For more details on the computations which lead to the representation of Y_n given above, the reader is referred to [19], the classical reference on Bessel functions.

The asymptotic behaviour of $Y_n(x)$ as $x \to 0$ is given by

$$Y_n(x) \sim \begin{cases} \dfrac{2}{\pi} \log \dfrac{x}{2}, & n = 0 \\[2mm] -\dfrac{(n-1)!}{\pi} \left(\dfrac{x}{2}\right)^{-n}, & n \in \mathbb{N}, \end{cases} \tag{5.17}$$

where $f(x) \sim g(x)$ as $x \to c$ means $f(x)$ approaches $g(x)$ asymptotically as $x \to c$, that is, $f(x)/g(x) \to 1$ as $x \to c$. Thus $Y_n(x)$ is unbounded in the neighborhood of $x = 0$. As $x \to 0$,

$$Y_0(x) \sim \frac{2}{\pi} \log x, \quad Y_1(x) \sim -\frac{2}{\pi}\frac{1}{x}, \quad Y_2(x) \sim \frac{4}{\pi}\frac{1}{x^2}, \quad \cdots .$$

In view of Theorem 5.3 and the Sturm separation theorem, we can conclude that Y_n has an infinite sequence of zeros in $(0, \infty)$ which alternate with the zeros of J_n (Fig. 5.3).

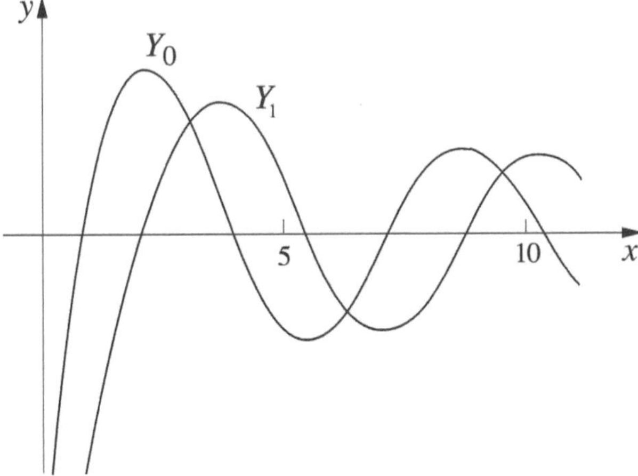

Fig. 5.3 Bessel functions Y_0 and Y_1

Exercises

5.17 Verify the asymptotic behaviour of Y_0 and Y_1 near $x = 0$ as expressed by (5.17).

5.18 Prove that

$$\frac{d}{dx}[x^\nu Y_\nu(x)] = x^\nu Y_{\nu-1}(x).$$

5.19 Prove that

$$\frac{d}{dx}[x^{-\nu} Y_\nu(x)] = -x^{-\nu} Y_{\nu+1}(x).$$

5.20 Prove that $Y_{-n}(x) = (-1)^n Y_n(x)$ for all $n \in \mathbb{N}_0$.

5.21 The *modified Bessel function of the first kind I_ν* is defined on $(0, \infty)$ by

$$I_\nu(x) = i^{-\nu} J_\nu(ix), \quad \nu \geq 0,$$

where $i = \sqrt{-1}$. Show that I_ν satisfies the equation

$$x^2 y'' + x y' - (x^2 + \nu^2) y = 0. \tag{5.18}$$

5.22 Based on the definition of I_ν in Exercise 5.21, show that I_ν is a real function represented by the series (see Fig. 5.4)

$$I_\nu(x) = \sum_{m=0}^{\infty} \frac{1}{m! \, \Gamma(m + \nu + 1)} \left(\frac{x}{2}\right)^{2m+\nu}.$$

5.23 Prove that $I_\nu(x) \neq 0$ for any $x > 0$, $\nu > -1$, and that $I_{-n}(x) = I_n(x)$ for all $n \in \mathbb{N}$.

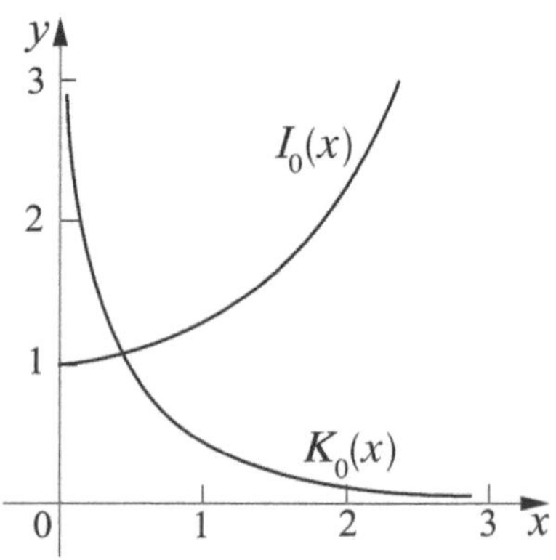

Fig. 5.4 Modified Bessel functions I_0 and K_0

5.24 Show that the *modified Bessel function of the second kind* (Fig. 5.4)

$$K_\nu(x) = \frac{\pi}{2 \sin \nu\pi} [I_{-\nu}(x) - I_\nu(x)]$$

also satisfies Eq. (5.18).

5.4   Integral Form of the Bessel Function J_n

We first prove that the generating function of J_n is

$$e^{x(z-1/z)/2} = \sum_{n=-\infty}^{\infty} J_n(x)z^n, \quad z \neq 0. \tag{5.19}$$

This can be seen by noting that

$$e^{xz/2} = \sum_{j=0}^{\infty} \frac{z^j}{j!} \left(\frac{x}{2}\right)^j,$$

$$e^{-x/2z} = \sum_{k=0}^{\infty} \frac{(-1)^k}{k!z^k} \left(\frac{x}{2}\right)^k,$$

and that these two series are absolutely convergent for all x in $\mathbb{R}$, $z \neq 0$, hence their product is the double series

$$e^{x(z-1/z)/2} = \sum_{j=0}^{\infty} \sum_{k=0}^{\infty} \frac{(-1)^k}{j!k!} \left(\frac{x}{2}\right)^{j+k} z^{j-k}.$$

Setting $j - k = n$, and recalling that $1/(k+n)! = 1/\Gamma(k+n+1) = 0$ when $k + n < 0$, we obtain

$$e^{x(z-1/z)/2} = \sum_{n=-\infty}^{\infty} \left[\sum_{k=0}^{\infty} \frac{(-1)^k}{k!(k+n)!} \left(\frac{x}{2}\right)^{2k+n} \right] z^n$$

$$= \sum_{n=-\infty}^{\infty} J_n(x)z^n,$$

which proves (5.19). The substitution $z = e^{i\theta}$ now gives

$$\frac{1}{2}\left(z - \frac{1}{z}\right) = i \sin\theta,$$

hence

$$e^{ix\sin\theta} = \sum_{n=-\infty}^{\infty} J_n(x)e^{in\theta}. \tag{5.20}$$

The function $e^{ix\sin\theta}$ is periodic with period 2π and satisfies the conditions of Theorem 3.9, hence the right-hand side of (5.20) represents its Fourier series expansion in exponential form, and $J_n(x)$ are the Fourier coefficients in the expansion. Therefore

$$\begin{aligned}
J_n(x) &= \frac{1}{2\pi} \int_{-\pi}^{\pi} e^{ix\sin\theta} e^{-in\theta} d\theta \\
&= \frac{1}{2\pi} \int_{-\pi}^{\pi} e^{i(x\sin\theta - n\theta)} d\theta \\
&= \frac{1}{2\pi} \int_{-\pi}^{\pi} \cos(x\sin\theta - n\theta)d\theta, \text{ since } J_n(x) \text{ is real,} \\
&= \frac{1}{\pi} \int_{0}^{\pi} \cos(x\sin\theta - n\theta)d\theta, \quad n \in \mathbb{N}_0, \tag{5.21}
\end{aligned}$$

which is the principal integral representation of J_n. The formula (5.21) immediately gives an upper bound on J_n,

$$|J_n(x)| \leq \frac{1}{\pi} \int_{0}^{\pi} d\theta = 1 \text{ for all } n \in \mathbb{N}_0,$$

a result we would not have been able to infer directly from the series definition. With $J_n(0) = \frac{1}{\pi} \int_0^\pi \cos n\theta d\theta$, it also confirms that $J_0(0) = 1$ and $J_n(0) = 0$ for all $n \geq 1$.

Going back to Eq. (5.20) and equating the real and the imaginary parts of both sides, we have

$$\cos(x\sin\theta) = \sum_{n=-\infty}^{\infty} J_n(x)\cos n\theta,$$

$$\sin(x\sin\theta) = \sum_{n=-\infty}^{\infty} J_n(x)\sin n\theta.$$

With $J_{-n}(x) = (-1)^n J_n(x)$, this implies

$$\cos(x\sin\theta) = J_0(x) + 2\sum_{m=1}^{\infty} J_{2m}(x)\cos 2m\theta, \tag{5.22}$$

$$\sin(x \sin\theta) = 2 \sum_{m=1}^{\infty} J_{2m-1}(x)\sin(2m-1)\theta. \tag{5.23}$$

Now Eqs. (5.22) and (5.23) are, respectively, the Fourier series expansions of the even function $\cos(x \sin\theta)$ and the odd function $\sin(x \sin\theta)$. Hence we arrive at the following pair of integral formulas for their coefficients,

$$J_{2m}(x) = \frac{1}{\pi} \int_0^{\pi} \cos(x \sin\theta) \cos 2m\theta \, d\theta, \quad m \in \mathbb{N}_0, \tag{5.24}$$

$$J_{2m-1}(x) = \frac{1}{\pi} \int_0^{\pi} \sin(x \sin\theta) \sin(2m-1)\theta \, d\theta, \quad m \in \mathbb{N}. \tag{5.25}$$

Exercises

5.25 Show that

$$J_n'(x) = \frac{1}{\pi} \int_0^{\pi} \sin\theta \sin(n\theta - x \sin\theta)d\theta,$$

then use induction, or Eq. (5.20), to prove that

$$J_n^{(k)} = \frac{1}{\pi} \int_0^{\pi} \sin^k\theta \cos(n\theta - x \sin\theta - k\pi/2)d\theta, \quad k \in \mathbb{N}.$$

5.26 Use the result of Exercise 5.25 to prove that $\left|J_n^{(k)}(x)\right| \le 1$ for all $n, k \in \mathbb{N}_0$.

5.27 Prove that

(a) $J_0(x) + 2 \sum_{m=1}^{\infty} J_{2m}(x) = 1.$

(b) $J_0(x) + 2 \sum_{m=1}^{\infty} (-1)^m J_{2m}(x) = \cos x.$

(c) $2 \sum_{m=0}^{\infty} (-1)^m J_{2m+1}(x) = \sin x.$

(d) $\sum_{m=1}^{\infty} (2m-1) J_{2m-1}(x) = x/2.$

5.28 Use Parseval's relation to prove the identity

$$J_0^2(x) + 2 \sum_{n=1}^{\infty} J_n^2(x) = 1.$$

Note that this implies $|J_0(x)| \le 1$ and $|J_n(x)| \le 1/\sqrt{2}$, $n \in \mathbb{N}$.

5.29 Show how Eqs. (5.22) and (5.23) imply (5.24) and (5.25).
5.30 Prove that $\lim_{n\to\infty} J_n(x) = 0$ for all $x \in \mathbb{R}$.

5.5 Orthogonality Property

After division by x Bessel's equation takes the form

$$xy'' + y' + (x - \frac{v^2}{x})y = 0, \tag{5.26}$$

where the differential operator

$$L = \frac{d}{dx}\left(x\frac{d}{dx}\right) - \frac{v^2}{x}$$

is formally self-adjoint, with $p(x) = x$ and $r(x) = -v^2/x$ in the standard form (2.33). Comparison with Eq. (2.34) shows that $\rho(x) = x$ is the weight function, and the eigenvalue parameter does not appear explicitly in Eq. (5.26). But we can introduce a parameter μ through the change of variables

$$x \mapsto \mu x, \quad y(x) \mapsto y(\mu x) = u(x).$$

Differentiating with respect to x,

$$u'(x) = \mu y'(\mu x)$$

$$u''(x) = \mu^2 y''(\mu x).$$

Under this transformation, Eq. (5.26) becomes

$$xu'' + u' + \left(\mu^2 x - \frac{v^2}{x}\right)u = 0, \tag{5.27}$$

where the eigenvalue parameter is now $\lambda = \mu^2$. Equations (5.26) and (5.27) are equivalent provided $\mu \neq 0$.

If Eq. (5.27) is given on the interval (a, b), where $0 \leq a < b < \infty$, then we can impose the homogeneous, separated boundary conditions

$$\alpha_1 u(a) + \alpha_2 u'(a) = 0, \quad \beta_1 u(b) + \beta_2 u'(b) = 0,$$

to complete the set-up of the SL problem for Bessel's equation. The general solution is given by

$$c_\mu J_v(\mu x) + d_\mu Y_v(\mu x),$$

where μ, c_μ, and d_μ are chosen so that the boundary conditions are satisfied. The details are generally quite tedious, and we simplify things by taking $a = 0$. We further restrict ν to the non-negative integers, since the assumption $n \in \mathbb{N}_0$ is more useful from the point of view of physical applications.

Suppose, therefore, that Eq. (5.27) is given on the interval $(0, b)$. Since $p(0) = 0$, no boundary condition is needed at $x = 0$, except that $\lim_{x \to 0+} u(x)$ exist. At $x = b$ we have

$$\beta_1 u(b) + \beta_2 u'(b) = 0. \tag{5.28}$$

The pair of Eqs. (5.27) and (5.28) pose a singular SL eigenvalue problem and, based on an extension of the theory developed in Chap. 2, it has an orthogonal set of solutions which is complete in $\mathcal{L}_\rho^2(0, b)$. The general solution of Eq. (5.27) is now given by

$$c_n J_n(\mu x) + d_n Y_n(\mu x).$$

The condition that $\lim_{x \to 0+} u(x)$ exist forces the coefficient of Y_n to vanish, and we are left with $J_n(\mu x)$ as the only admissible solution.

Let us start with the special case of Eq. (5.28) when $\beta_2 = 0$, that is,

$$u_n(b) = 0. \tag{5.29}$$

Applying this condition to the solution $J_n(\mu x)$ gives

$$J_n(\mu b) = 0, \quad n \in \mathbb{N}_0. \tag{5.30}$$

We have already determined in Theorem 5.3 that, for each n, the roots of Eq. (5.30) in $(0, \infty)$ form an infinite increasing sequence which tends to ∞,

$$\xi_{n1} < \xi_{n2} < \xi_{n3} < \cdots .$$

The solutions of Eq. (5.30) are therefore given by

$$\mu_k b = \xi_{nk}, \quad k \in \mathbb{N},$$

and the eigenvalues of Eq. (5.27) are $\lambda_k = \mu_k^2 = \left(\xi_{nk}/b\right)^2$.

The first zero $\xi_{n0} = 0$ of the function J_n, for $n \geq 1$, does not determine an eigenvalue because the corresponding solution is

$$J_n(\mu_0 x) = J_n(0) = 0 \text{ for all } n \in \mathbb{N},$$

which is not an eigenfunction, and $J_0(0) \neq 0$. The sequence of eigenvalues of the system (5.27), (5.28) is therefore

$$0 < \lambda_1 = \mu_1^2 < \lambda_2 = \mu_2^2 < \lambda_3 = \mu_3^2 < \cdots ,$$

and its corresponding sequence of eigenfunctions is

$$J_n(\mu_1 x), \ J_n(\mu_2 x), \ J_n(\mu_3 x), \ \cdots .$$

For each $n \in \mathbb{N}_0$, the sequence $(J_n(\mu_k x) : k \in \mathbb{N})$ is necessarily orthogonal (as they are associated with distinct eigenvalues) and complete in $\mathcal{L}_x^2(0, b)$, so

$$\langle J_n(\mu_j x), J_n(\mu_k x)\rangle_x = \int_0^b J_n(\mu_j x) J_n(\mu_k x) x \, dx = 0 \quad \text{for all } j \neq k, \qquad (5.31)$$

and, for any $f \in \mathcal{L}_x^2(0, b)$ and any $n \in \mathbb{N}_0$, we can represent f by the *Fourier-Bessel series*

$$f(x) = \sum_{k=1}^{\infty} \frac{\langle f(x), J_n(\mu_k x)\rangle_x}{\left\| J_n(\mu_k x) \right\|_x^2} J_n(\mu_k x), \qquad (5.32)$$

the latter equality being in $\mathcal{L}_x^2(0, b)$. If f is piecewise smooth on $(0, b)$ Eq. (5.32) holds pointwise as well, provided $f(x)$ is defined as $\frac{1}{2}[f(x^+) + f(x^-)]$ at the points of discontinuity of f.

The orthogonality relation (5.31) is of course a consequence of the self-adjointness of the differential operator L with the given boundary condition, and we will not attempt to verify it by direct computation. But we need to determine $\left\| J_n(\mu_k x) \right\|_\rho^2$ in order to use the Fourier-Bessel series expansion (5.32). Multiplying Eq. (5.27) by $2xu'$, we obtain

$$2xu'(xu')' + (\mu^2 x^2 - v^2)2uu' = 0$$

$$[(xu')^2]' + (\mu^2 x^2 - v^2)(u^2)' = 0.$$

Integrating this last equation over $(0, b)$,

$$(xu')^2 \Big|_0^b + \mu^2 \left[x^2 u^2 \Big|_0^b - 2 \int_0^b xu^2 dx \right] - v^2 u^2 \Big|_0^b = 0,$$

$$\Rightarrow \quad \|u\|_x^2 = \frac{1}{2\mu^2} \left[(\mu x u)^2 + (xu')^2 - v^2 u^2 \right] \Big|_0^b .$$

With $v = n \in \mathbb{N}_0$, $u(x) = J_n(\mu x)$, $u'(x) = \mu J_n'(\mu x)$, and $\mu > 0$, we therefore have

$$\| J_n(\mu x) \|_x^2 = \int_0^b J_n^2(\mu x) x\, dx$$

$$= \frac{1}{2\mu^2} \left[(\mu^2 x^2 - n^2) J_n^2(\mu x) + \mu^2 x^2 J_n'^2(\mu x) \right] \Big|_0^b .$$

Since $n^2 J_n(0) = 0$ for all $n \in \mathbb{N}_0$,

$$\| J_n(\mu x) \|_x^2 = \frac{b^2}{2} \left[J_n'(\mu b) \right]^2 + \frac{\mu^2 b^2 - n^2}{2\mu^2} J_n^2(\mu b). \tag{5.33}$$

When $\mu = \mu_k$ the last term drops out, by (5.30), and

$$\left\| J_n(\mu_k x) \right\|_x^2 = \frac{b^2}{2} [J_n'(\mu_k b)]^2 .$$

Using the result of Exercise 5.9,

$$J_n'(\mu_k b) = \frac{1}{\mu_k b} [n J_n(\mu_k b) - \mu_k b J_{n+1}(\mu_k b)] = -J_{n+1}(\mu_k b),$$

we finally obtain

$$\left\| J_n(\mu_k x) \right\|_x^2 = \frac{b^2}{2} J_{n+1}^2(\mu_k b). \tag{5.34}$$

Example 5.5 To expand the function

$$f(x) = \begin{cases} 1, \ 0 \le x < 2 \\ 0, \ 2 < x \le 4 \end{cases}$$

in a Fourier-Bessel series under the condition $J_0(4\mu) = 0$, we first evaluate the coefficients in the series :

$$\langle f(x), J_0(\mu_k x) \rangle_x = \int_0^4 f(x) J_0(\mu_k x) x\, dx$$

$$= \int_0^2 J_0(\mu_k x) x\, dx$$

$$= \frac{1}{\mu_k^2} \int_0^{2\mu_k} J_0(y) y\, dy$$

$$= \frac{2}{\mu_k} J_1(2\mu_k),$$

where we used the result of Example 5.4 in the last equality, $4\mu_k$ being the zeros of J_0. From the formula (5.34) we get

$$\left\| J_0(\mu_k x) \right\|_x^2 = 8 J_1^2(4\mu_k).$$

Now, by (5.32),

$$f(x) = \frac{1}{4} \sum_{k=1}^{\infty} \frac{J_1(2\mu_k)}{\mu_k J_1^2(4\mu_k)} J_0(\mu_k x), \ 0 < x < 4.$$

Observe that by setting $x = 1$ in this equation, which is a point of continuity for f, we arrive at the identity

$$\sum_{k=1}^{\infty} \frac{J_1(2\mu_k) J_0(\mu_k)}{\mu_k J_1^2(4\mu_k)} = 4.$$

At the point of discontinuity $x = 2$, we have

$$\sum_{k=1}^{\infty} \frac{J_1(2\mu_k) J_0(2\mu_k)}{\mu_k J_1^2(4\mu_k)} = 4 \frac{f(2^+) + f(2^-)}{2} = 2.$$

The more general condition (5.28)

$$\beta_1 u(b) + \beta_2 u'(b) = 0$$

can, in principle, be handled in a similar fashion, though it requires more work. If $\beta_1 \neq 0$ and $\beta_2 \neq 0$, we can assume, without loss of generality, that $\beta_2 = 1$. The solution $u(x) = J_n(\mu x)$ must then satisfy

$$\beta_1 J_n(\mu b) + \mu J_n'(\mu b) = 0. \tag{5.35}$$

The positive roots of this equation determine the eigenvalues μ_k^2 of the problem. The corresponding eigenfunctions are $u_k(x) = J_n(\mu_k x)$. The norm of each eigenfunction is determined by substituting into (5.33) to obtain

$$\left\| J_n(\mu_k x) \right\|_x^2 = \frac{b^2}{2} \left[J_n'(\mu_k b) \right]^2 + \frac{\mu_k^2 b^2 - n^2}{2\mu_k^2} J_n^2(\mu_k b)$$

$$= \frac{1}{2\mu_k^2} \left[\beta_1^2 b^2 + \mu_k^2 b^2 - n^2 \right] J_n^2(\mu_k b). \tag{5.36}$$

To solve Eq. (5.35) for μ is, in general, not a simple matter. We should recall that J_n and J_n' cannot have a common zero, because the solution to the initial-value problem for Bessel's equation is unique, so one would most likely have to resort to numerical methods to determine μ_k. But if $\beta_1 = 0$, that is, if the boundary condition at b is

$$u'(b) = 0, \tag{5.37}$$

then the eigenvalues are determined by the roots of the equation $\mu J_n'(\mu b) = 0$. The first is $\mu_0 = 0$, corresponding to the eigenvalue

$$\lambda_0 = 0,$$

and the eigenfunction

$$u_0(x) = J_0(0) = 1.$$

Note that this function solves Eq. (5.27) but not (5.26). As mentioned earlier, these two equations are not equivalent if $\mu = 0$.

With $n = 0$, the other values of μ_k must satisfy $J_0'(\mu_k b) = 0$ or, equivalently, $J_1(\mu_k b) = 0$. Thus the sequence

$$\mu_k = \xi_{1k}/b, \quad k = 1, 2, 3, \cdots,$$

where ξ_{1k} are the positive zeros of J_1, yield the remaining eigenvalues of the problem,

$$\lambda_k = \frac{\xi_{1k}^2}{b^2}, \quad k \in \mathbb{N},$$

corresponding to the eigenfunctions

$$u_k(x) = J_0(\mu_k x), \quad k \in \mathbb{N}.$$

Under the boundary condition (5.37), if n is a positive integer, then 0 is not an eigenvalue, and we obtain the eigenvalues $\lambda_k = \mu_k^2$ by solving

$$J_n'(\mu_k b) = 0.$$

The corresponding eigenfunctions are then $u_k(x) = J_n(\mu_k x)$. In any case, whether n is 0 or positive, the norm of J_n is calculated from the formula (5.36) with $\beta_1 = 0$.

Example 5.6 In $\mathbb{R}^3$ the cylindrical coordinates (r, θ, z) are related to the cartesian coordinates (x, y, z) by

$$x = r\cos\theta, \quad y = r\sin\theta,$$

$$r = \sqrt{x^2 + y^2},$$

where $0 \le r < \infty$ and $-\pi < \theta \le \pi$. The Laplacian operator in cylindrical coordinates has the form

$$\Delta = \frac{\partial^2}{\partial r^2} + \frac{1}{r}\frac{\partial}{\partial r} + \frac{1}{r^2}\frac{\partial^2}{\partial \theta^2} + \frac{\partial^2}{\partial z^2}.$$

Given the cylindrical region

$$\Omega = \{(r, \theta, z) : 0 \le r < b, -\pi < \theta \le \pi, 0 < z < h\},$$

we seek a potential function u which satisfies Laplace's equation

$$\Delta u = u_{rr} + r^{-1}u_r + r^{-2}u_{\theta\theta} + u_{zz} = 0$$

in Ω, and assumes the following values on the boundary of Ω :

$$u(r, \theta, 0) = 0, \quad 0 \le r \le b, -\pi < \theta \le \pi,$$
$$u(b, \theta, z) = 0, \quad -\pi < \theta \le \pi, 0 \le z \le h,$$
$$u(r, \theta, h) = f(r), \quad 0 \le r \le b, -\pi < \theta \le \pi.$$

Solution u is independent of θ on the boundary, so we can assume, by symmetry, that it is also independent of θ inside Ω. Using separation of variables, let

$$u(r, z) = v(r)w(z), \quad 0 \le r \le b,$$

and substitute into Laplace's equation to obtain

$$v''(r)w(z) + r^{-1}v'(r)w(z) + v(r)w''(z) = 0.$$

This leads to the pair of equations

$$r^2 v''(r) + r v'(r) + \mu^2 r^2 v(r) = 0 \tag{5.38}$$

$$w''(z) - \mu^2(z) = 0, \tag{5.39}$$

where the separation constant was assumed to be $-\mu^2$. Equation (5.38) is Bessel's equation with $\nu = 0$ and $\rho = r$, whose continuous solution in $0 \leq r \leq b$ is $J_0(\mu r)$. Under the boundary condition $v(b) = 0$, the eigenfunctions are $J_0(\mu_k r)$ corresponding to the eigenvalues $\lambda_k = \mu_k^2 = (\xi_{0k}/b)^2$. As before, $\xi_{01}, \xi_{02}, \xi_{03}, \cdots$ are the positive zeros of J_0.

The corresponding eigenfunctions of Eq. (5.39) under the condition $w(0) = 0$ are $\sinh(\mu_k z)$. Hence the sequence of eigenfunctions of the original problem are

$$u_k(r, z) = J_0\left(\mu_k r\right) \sinh\left(\mu_k z\right), \quad k \in \mathbb{N},$$

and the general solution is

$$u(r, z) = \sum_{k=1}^{\infty} c_k u_k(r, z).$$

Now the condition $u(r, h) = f(r)$ implies

$$f(r) = \sum_{k=1}^{\infty} c_k \sinh\left(\mu_k h\right) J_0\left(\mu_k r\right), \tag{5.40}$$

which is the Fourier-Bessel series expansion of the function f. The coefficients in the expansion are given by

$$c_k \sinh\left(\mu_k h\right) = \frac{1}{\left\| J_0(\mu_k r) \right\|_r^2} \int_0^b f(r) J_0(\mu_k r) r \, dr,$$

and this completely determines c_k for all positive integers k.

This is an example of what is known as a *Dirichlet problem* [13]. The general Dirichlet problem is to determine the function u in a domain Ω which satisfies Laplace's equation $\Delta u = 0$ in Ω and $u = f$ on the boundary $\partial\Omega$, where f is a given function defined on $\partial\Omega$. When Ω is bounded, its boundary sufficiently smooth, and the function f is continuous, the problem is known to have a unique solution in $C^2(\Omega)$. The problem of solving $\Delta u = 0$ in the sphere $0 \leq r < R$ under a given boundary condition on $r = R$, which was treated in Sect. 4.4.1, is another example of a Drichlet problem. There the solution was expressed in terms of Legendre polynomials. Bessel functions often appear in the solution of Laplace's equation in cylindrical coordinates, as we saw in this last example, hence they are sometimes called *cylinder functions*. It would seem that, not only the differential equation, but also the geometry of the domain Ω often determines the appropriate eigenfunction expansion of the solution.

Exercises

5.31 Determine the Fourier-Bessel series expansion $\sum c_k J_0(\mu_k x)$ on $[0, b]$, where μ_k are the positive roots of the equation $J_0(\mu b) = 0$, for each of the following functions:

(a) $f(x) = 1$
(b) $f(x) = x$
(c) $f(x) = x^2$
(d) $f(x) = b^2 - x^2$
(e) $f(x) = \begin{cases} 1, & 0 < x < b/2 \\ 1/2, & x = b/2 \\ 0, & b/2 < x < b. \end{cases}$

5.32 Expand the function $f(x) = 1$ on $[0, 1]$ in terms of $J_0(\mu_k x)$, where μ_k are the non-negative zeros of J_0'.

5.33 Expand $f(x) = x$ on $[0, 1]$ in terms of $J_1(\mu_k x)$, where μ_k are the positive zeros of J_1.

5.34 For any positive integer n, expand $f(x) = x^n$ on $[0, 1]$ in terms of $J_n(\mu_k x)$, where μ_k are the positive zeros of J_n'.

5.35 Determine the coefficients in the series $\sum c_k J_1(\mu_k x)$ which represents the function

$$f(x) = \begin{cases} x, & 0 \leq x \leq 1 \\ 0, & 1 < x \leq 2 \end{cases}$$

on $[0, 2]$, where μ_k are the positive zeros of $J_1'(2\mu)$. Is the representation pointwise on $[0, 2]$?

5.36 Show that 0 is an eigenvalue of Bessel's Equation (5.27) subject to the boundary condition (5.28) if, and only if, $\beta_1/\beta_2 = -\nu/b$, and that the corresponding eigenfunction is x^ν, where $\nu > 0$.

5.37 The heat equation on a circular flat plate is given in polar coordinates (r, θ) by

$$u_t = k \left(u_{rr} + \frac{1}{r} u_r + \frac{1}{r^2} u_{\theta\theta} \right).$$

Suppose that the temperature $u = u(r, t)$ does not depend on θ and that $0 \leq r < 1$. If the edge of the plate is held at zero temperature for all $t > 0$, use separation of variables to show that the temperature on the plate is given by

$$u(r, t) = \sum_{n=1}^{\infty} c_n e^{-\mu_n^2 k t} J_0(\mu_n r) \quad \text{for all } r \in [0, 1), t > 0, \qquad (5.41)$$

where μ_n are the positive zeros of J_0.

5.38 If the initial temperature on the plate in Exercise 5.37 is $u(r, 0) = f(r)$, determine the Fourier-Bessel coefficients in Eq. (5.41).

5.39 A thin elastic circular membrane vibrates transversally according to the wave equation

$$u_{tt} = c^2 \left(u_{rr} + \frac{1}{r} u_r \right), \quad 0 \leq r < R, t > 0.$$

If the boundary condition is $u(R, t) = 0$ for all $t > 0$, and the initial conditions are

$$u(r, 0) = f(r), \quad u_t(r, 0) = g(r), \quad 0 \leq r < R,$$

determine the form of the bounded solution $u(r, t)$ in terms of J_n for all $r \in [0, R)$ and $t > 0$.

Chapter 6
The Fourier Transformation

The underlying theme of the previous chapters was the Sturm-Liouville theory. The last three chapters show how the eigenfunctions of various SL problems serve as bases for $\mathcal{L}^2$, either through conventional Fourier series or its generalized version. In this chapter we introduce the Fourier integral as a limiting case of the classical Fourier series, and show how it serves, under certain conditions, as a method for representing non-periodic functions on $\mathbb{R}$ where the series approach does not apply. This chapter and the next are therefore concerned with extending the theory of Fourier series to non-periodic functions.

6.1 The Fourier Transform

Suppose $f : \mathbb{R} \to \mathbb{C}$ is an $\mathcal{L}^2$ function. Its restriction to $(-l, l)$ clearly lies in $\mathcal{L}^2(-l, l)$ for any $l > 0$. On the interval $(-l, l)$ we can always represent f by the Fourier series

$$f(x) = \sum_{n=-\infty}^{\infty} c_n e^{in\pi x/l}, \tag{6.1}$$

$$c_n = \frac{1}{2l} \int_{-l}^{l} f(x) e^{-in\pi x/l} dx, \quad n \in \mathbb{Z}. \tag{6.2}$$

Let $\Delta\xi = \pi/l$ and $\xi_n = n\Delta\xi = n\pi/l$. The pair of Eqs. (6.1) and (6.2) then take the form

$$f(x) = \frac{1}{2\pi} \sum_{-\infty}^{\infty} C(\xi_n) e^{i\xi_n x} \Delta\xi, \tag{6.3}$$

$$C(\xi_n) = 2lc_n = \int_{-l}^{l} f(x)e^{-i\xi_n x}dx. \tag{6.4}$$

If we now let $l \to \infty$, that is, if we allow the period $(-l, l)$ to expand to $\mathbb{R}$ and f loses its periodicity, then the discrete variable ξ_n will behave more like a real variable ξ, and the formula (6.4) will approach the form

$$C(\xi) = \int_{-\infty}^{\infty} f(x)e^{-i\xi x}dx. \tag{6.5}$$

The right-hand side of (6.3), on the other hand, looks very much like a Riemann sum which, in the limit as $l \to \infty$, approaches the integral

$$f(x) = \frac{1}{2\pi} \int_{-\infty}^{\infty} C(\xi)e^{ix\xi}d\xi. \tag{6.6}$$

Thus the Fourier coefficients c_n are transformed to the function $C(\xi)$, the transform of f, and the Fourier series (6.1), which represents f on $(-l, l)$, is replaced by the integral (6.6) which, presumably, represents the function f on $(-\infty, \infty)$.

The procedure described above is, of course, not intended to be a "proof" of the validity of the formulas (6.5) and (6.6). The integral in (6.5) may not even exist. It is meant to be a plausible argument for motivating the definition (to follow) of the Fourier transform, which can then be used to represent the (non-periodic) function f by the integral (6.6), in much the same way that the Fourier series were used in Chap. 3 to represent periodic functions.

For any real interval I, we use the symbol $\mathcal{L}^1(I)$ to denote the set of Lebesgue measurable functions $f : I \to \mathbb{C}$ such that

$$\int_I |f(x)|\, dx < \infty,$$

where the integral is defined, in the sense of Lebesgue, as the supremum of integrals of an increasng sequence $\{0 \le \varphi_n \le |f|\}$ of simple functions over I, as presented in section A.2 of Appendix A. The resulting class of Lebesgue integrable functions $\mathcal{L}^1(I)$ extends the proper Riemann integrable functions $\mathcal{R}(I)$. Though we have practically no occasion to deal with functions in $\mathcal{L}^1(I)$ which are not Riemann integrable, the integrals in this chapter will be considered Lebesgue integrals so that we can apply some results of the Lebesgue theory that cannot be justified in Riemann's theory of integration, as we shall presently show. In the meantime, here are some examples:

$$x^\alpha \in \mathcal{L}^1(0, 1) \Leftrightarrow \alpha > -1,$$

$$x^\alpha \in \mathcal{L}^1(1, \infty) \Leftrightarrow \alpha < -1,$$

$$\frac{\sin x}{x} \notin \mathcal{L}^1(1, \infty).$$

Definition 6.1 For any $f \in \mathcal{L}^1(\mathbb{R})$ the *Fourier transform* of f is the function $\hat{f} : \mathbb{R} \to \mathbb{C}$ defined by the integral

$$\hat{f}(\xi) = \int_{-\infty}^{\infty} f(x) e^{-i\xi x} dx. \tag{6.7}$$

We also use the symbol $\mathcal{F}(f)$ to denote the Fourier transform of f.

With $\left| e^{i\xi x} \right| = 1$ we clearly have

$$\left| \hat{f}(\xi) \right| \leq \int_{-\infty}^{\infty} |f(x)| \, dx < \infty,$$

that is, $\hat{f}$ is a bounded function on $\mathbb{R}$. By the linearity of the integral,

$$\mathcal{F}(c_1 f_1 + c_2 f_2) = c_1 \mathcal{F}(f_1) + c_2 \mathcal{F}(f_2)$$

for all $c_1, c_2 \in \mathbb{C}$ and all $f_1, f_2 \in \mathcal{L}^1(\mathbb{R})$, which means that the Fourier transformation $\mathcal{F} : f \to \hat{f}$ is linear.

Example 6.2 For any positive constant a, let

$$f_a(x) = \begin{cases} 1, & |x| \leq a \\ 0, & |x| > a. \end{cases}$$

Then

$$\hat{f}_a(\xi) = \int_{-a}^{a} e^{-i\xi x} dx = \frac{1}{-i\xi} \left(e^{-i\xi a} - e^{i\xi a} \right) = \frac{2}{\xi} \sin a\xi.$$

Note that f_a is discontinuous at $x = \pm a$, but that $\hat{f}_a$ is continuous on $\mathbb{R}$ and tends to 0 as $|\xi| \to \infty$.

Example 6.3 In the case of the function $f(x) = e^{-|x|}$ we have

$$\hat{f}(\xi) = \int_{-\infty}^{0} e^{x} e^{-i\xi x} dx + \int_{0}^{\infty} e^{-x} e^{-i\xi x} dx$$

$$= \frac{1}{1 - i\xi} + \frac{1}{1 + i\xi}$$

$$= \frac{2}{1 + \xi^2}.$$

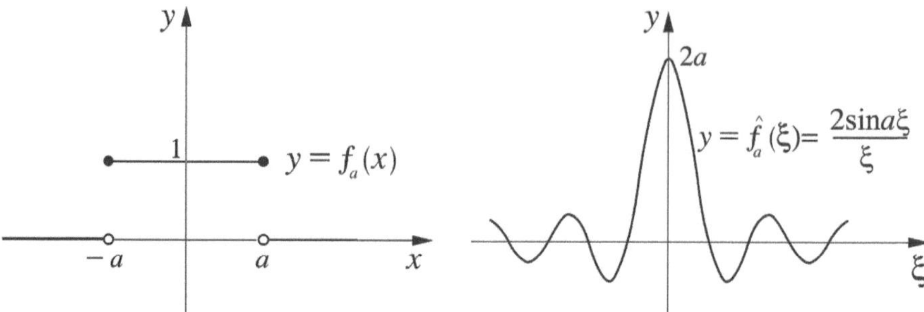

Fig. 6.1 Fourier transform of f_a

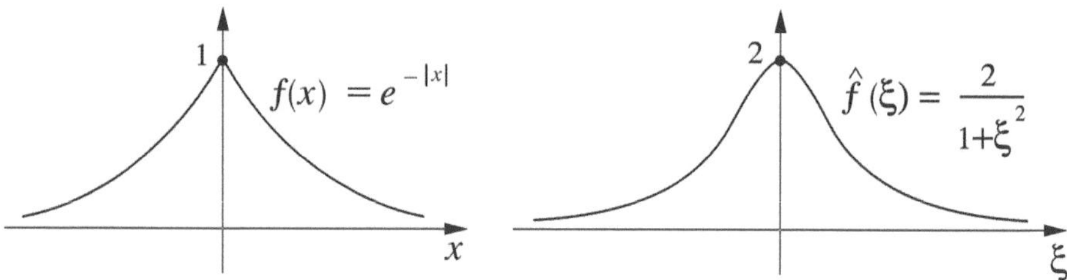

Fig. 6.2 Fourier transform of $e^{-|x|}$

Both Figs. 6.1 and 6.2 exhibit a smoother graph of the transform function $\hat{f}$ compared to that of f.

When $|f|$ is integrable over $\mathbb{R}$, i.e. when $f \in \mathcal{L}^1(\mathbb{R})$, we have seen that its Fourier transform $\hat{f}$ is bounded, but we can also prove that $\hat{f}$ is continuous. This result relies on a well-known theorem of real analysis, the Lebesgue dominated convergence theorem (see Appendix), which states the following.

Theorem 6.4 *Let $(f_n : n \in \mathbb{N})$ be a sequence of functions in $\mathcal{L}^1(I)$, where I is a real interval, and suppose $f_n \to f$ pointwise on I. If there is a positive function $g \in \mathcal{L}^1(I)$ such that*

$$|f_n(x)| \le g(x) \ \text{ for all } x \in I, \ n \in \mathbb{N},$$

then $f \in \mathcal{L}^1(I)$ and

$$\lim_{n \to \infty} \int_I f_n(x)dx = \int_I f(x)dx.$$

This is an important theorem in Lebesgue integration, where both the pointwise convergence of f_n and the inequality between $|f|$ and g need only be a.e (almost everywhere) in I, as would be expected since the functions in $\mathcal{L}^1$ are really classes of functions which are equal a.e. Note that the interval I is not assumed to be finite.

To prove that $\hat{f}$ is continuous, let ξ be any real number and suppose (ξ_n) is a sequence which converges to ξ. Since

$$\left|\hat{f}(\xi_n) - \hat{f}(\xi)\right| \leq \int_{-\infty}^{\infty} \left|e^{-i\xi_n x} - e^{-i\xi x}\right| |f(x)|\, dx,$$

$$\left|e^{-i\xi_n x} - e^{-i\xi x}\right| |f(x)| \leq 2|f(x)| \in \mathcal{L}^1(\mathbb{R}),$$

Theorem 6.4 implies

$$\lim_{n\to\infty} \left|\hat{f}(\xi_n) - \hat{f}(\xi)\right| \leq \lim_{n\to\infty} \int_{-\infty}^{\infty} \left|e^{-i\xi_n x} - e^{-i\xi x}\right| |f(x)|\, dx$$

$$= \int_{-\infty}^{\infty} \lim_{n\to\infty} \left|e^{-i\xi_n x} - e^{-i\xi x}\right| |f(x)|\, dx$$

$$= 0.$$

For the behaviour of $\hat{f}(\xi)$ as $|\xi| \to \infty$ we have the following result.

Lemma 6.5 *For any $f \in \mathcal{L}^1(\mathbb{R})$,*

$$\lim_{|\xi|\to\infty} \int_{-\infty}^{\infty} f(x)e^{i\xi x}\, dx = 0$$

The proof follows directly from the Riemann-Lebesgue lemma, as expressed by Equation (A.6),

$$\lim_{n\to\infty} \int_{-\infty}^{\infty} f(x)\cos nx\, dx = 0$$

for any $f \in \mathcal{L}^1(\mathbb{R})$, since $e^{i\xi x} = \cos \xi x + i \sin \xi x$ and $\sin \xi x = \cos(x - \pi/2)$.

We have therefore proved the following theorem, which sums up the main properties of the transform function $\hat{f}$.

Theorem 6.6 *For any $f \in \mathcal{L}^1(\mathbb{R})$, the Fourier transform*

$$\hat{f}(x) = \int_{-\infty}^{\infty} f(x)e^{-i\xi x}\, dx$$

is a bounded, continuous function on $\mathbb{R}$, and

$$\lim_{|\xi|\to\infty} \hat{f}(\xi) = 0. \tag{6.8}$$

Remark 6.7 Lemma 6.5 and Theorem 6.6 clearly remain valid if $e^{i\xi x}$ is replaced by either $\cos \xi x$ or $\sin \xi x$. Furthemore, Lemma 6.5 still holds if the interval of integration $(-\infty, \infty)$ is replaced by any subinterval.

We indicated in our heuristic introduction to this chapter that the Fourier transform $\hat{f}(\xi)$, denoted by $C(\xi)$ in Eq. (6.5), plays the role of the Fourier coefficients of the periodic function f in the limit as the period becomes unbounded. Hence the asymptotic behaviour $\hat{f}(\xi) \to 0$ as $\xi \to \pm\infty$ is in line with the behaviour of the Fourier coefficients c_n when $n \to \pm\infty$. But while Theorem 6.6 states some basic properties of $\hat{f}$, it says nothing about its role in representing f as suggested by Eq. (6.6), namely the validity of the inversion formula

$$f(x) = \frac{1}{2\pi} \int_{-\infty}^{\infty} \hat{f}(\xi) e^{ix\xi} \, d\xi. \tag{6.9}$$

The right-hand side of this equation is called the Fourier integral of f. This integral may not exist, even if the function $\hat{f}(\xi)$ tends to 0 as $|\xi| \to \infty$, unless it tends to 0 fast enough. Moreover, even if the Fourier integral exists, the equality (6.9) may not hold pointwise in $\mathbb{R}$ in spite of the continuity of $\hat{f}$. This will be the subject of the next section.

Exercises

6.1 Determine the Fourier transform of each of the following functions where it exists:

(a) $f_1(x) = \begin{cases} 1 - |x|, & |x| \le 1 \\ 0, & |x| > 1. \end{cases}$

(b) $f_2(x) = \begin{cases} 1, & -1 \le x \le 1 \\ 0, & \text{otherwise.} \end{cases}$

(c) $f_3(x) = \begin{cases} 0, & x = 0 \\ x^{-1}, & |x| > 0. \end{cases}$

6.2 Determine the Fourier transform functions in Exercise 6.1 which belong to $\mathcal{L}^1(\mathbb{R})$, $\mathcal{L}^2(\mathbb{R})$.

6.3 Show that if $\hat{f}(\xi)$ is the Fourier transform of $f(x)$, then for any $a \in \mathbb{R}$, $\hat{f}(\xi - a)$ is the Fourier transform of $e^{iax} f(x)$. Use this to determine the Fourier transform of $e^{-|x|+ia}$.

6.4 Determine the Fourier transform of e^{-x^2}, $x \in \mathbb{R}$. Hint: use the definition and complete the square on the exponent, then convert to a double integral.

6.5 Investigate the convergence and continuity properties of the integral

$$F\left(\xi\right) = \int_0^\infty \frac{\sin \xi x}{x} dx$$

over $\mathbb{R}$.

6.6 Show that, for any positive number ξ,

$$\int_0^\infty x^n e^{-\xi x} dx = \frac{n!}{\xi^{n+1}}, \quad n \in \mathbb{N}_0.$$

Note that if we set $\xi = 1$ in this equation, we arrive at the familiar relation $\Gamma(n+1) = n!$.

6.7 Show that $\mathcal{F}\left(f\left(cx\right)\right) = c^{-1}\hat{f}\left(c^{-1}\xi\right)$ for any $c \in \mathbb{R} \setminus \{0\}$.

6.8 Given $f\left(x\right) = \cos x$ on $[-\pi, \pi]$, and 0 otherwise, find $\hat{f}\left(x\right)$ and determine whether it lies in $\mathcal{L}^1\left(\mathbb{R}\right)$.

6.9 Let f and g be piecewise smooth functions on (a, b), and suppose that $x_1, \cdots, x_n$ are their points of discontinuity in ascending order. Prove the following generalization of the formula for integration by parts:

$$\int_a^b f(x)g'(x)dx = f(b^-)g(b^-) - f(a^+)g(a^+)$$

$$+ \sum_{k=1}^n [f(x_k^-)g(x_k^-) - f(x_k^+)g(x_k^+)] - \int_a^b f'(x)g(x)dx.$$

6.2   The Fourier Integral

The main result of this section is Theorem 6.10, which establishes the inversion formula for the Fourier transform. The proof of the theorem relies on evaluating the improper integral

$$\int_0^\infty \frac{\sin x}{x} dx,$$

otherwise known as *Dirichlet's integral*. To show that this integral exists, we write

$$\int_0^\infty \frac{\sin x}{x} dx = \int_0^1 \frac{\sin x}{x} dx + \lim_{b \to \infty} \int_1^b \frac{\sin x}{x} dx. \tag{6.10}$$

The function $\sin x / x$ is continuous and bounded on $(0, 1]$, where it satisfies $0 \leq \sin x / x \leq 1$, so the first integral on the right-hand side of (6.10) exists. Using integration by parts in the second integral,

$$\int_1^b \frac{\sin x}{x}\,dx = \cos 1 - \frac{\cos b}{b} - \int_1^b \frac{\cos x}{x^2}\,dx,$$

and noting that

$$\left|\int_1^b \frac{\cos x}{x^2}\,dx\right| \leq \int_1^b \left|\frac{\cos x}{x^2}\right|\,dx$$

$$\leq \int_1^b \frac{1}{x^2}\,dx$$

$$\leq 1 - \frac{1}{b},$$

we see that $\lim_{b\to\infty}\int_1^b (\cos x/x^2)\,dx$ also exists. Hence the integral $\int_1^\infty (\sin x/x)\,dx$ is convergent (see Exercise 1.29 for another approach). The integrand is shown graphically in Fig. 6.3. Now that we know Dirichlet's integral exists, it remains to determine its value.

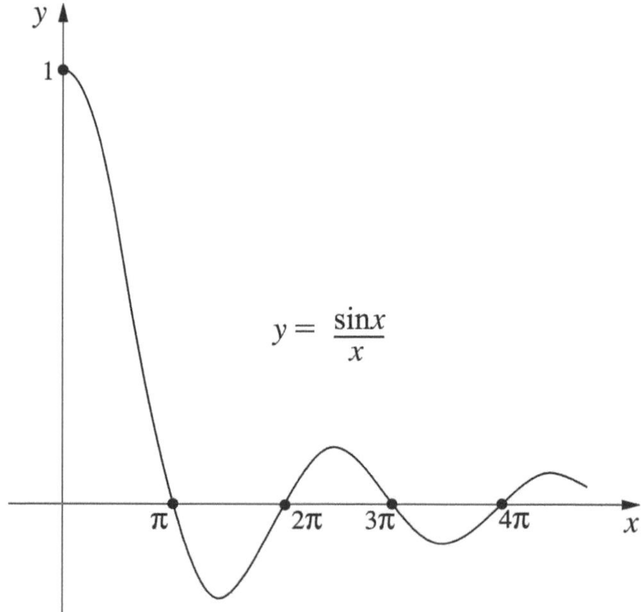

Fig. 6.3 The function $\sin x/x$

Lemma 6.8

$$\int_0^\infty \frac{\sin x}{x}\,dx = \frac{\pi}{2}. \tag{6.11}$$

Proof Let

$$f(x) = \begin{cases} \dfrac{1}{x} - \dfrac{1}{2\sin\frac{1}{2}x}, & 0 < x \le \pi \\[2mm] 0, & x = 0. \end{cases}$$

It is a simple matter to check that f and its derivative are both continuous on $[0, \pi]$. By Lemma 6.5,

$$\lim_{\xi \to \infty} \int_0^\pi f(x)\sin\xi x\,dx = 0. \tag{6.12}$$

Therefore

$$\begin{aligned} \int_0^\infty \frac{\sin x}{x}\,dx &= \lim_{\xi \to \infty} \int_0^{\pi\xi} \frac{\sin x}{x}\,dx \\[2mm] &= \lim_{\xi \to \infty} \int_0^\pi \frac{\sin\xi x}{x}\,dx \\[2mm] &= \lim_{\xi \to \infty} \frac{1}{2}\int_0^\pi \frac{\sin\xi x}{\sin\frac{1}{2}x}\,dx \quad \text{by (6.12)} \\[2mm] &= \lim_{n \to \infty} \frac{1}{2}\int_0^\pi \frac{\sin(n+\frac{1}{2})x}{\sin\frac{1}{2}x}\,dx. \end{aligned}$$

Going back to the definition of the Dirichlet kernel

$$D_n(x) = \frac{1}{2\pi}\sum_{k=-n}^n e^{ikx},$$

and using Lemma 3.8,

$$D_n(x) = \frac{1}{2\pi}\frac{\sin(n+\frac{1}{2})x}{\sin\frac{1}{2}x},$$

we conclude that

$$\int_0^\infty \frac{\sin x}{x}\,dx = \lim_{n\to\infty} \pi \int_0^\pi D_n(x)\,dx = \frac{\pi}{2},$$

by Eq. (3.17).   □

Although the function

$$\frac{\sin \xi x}{x}$$

is continuous with respect to both $x > 0$ and $\xi \in \mathbb{R}$, the function defined by the improper integral

$$K(\xi) = \int_0^\infty \frac{\sin \xi x}{x}\,dx$$

is not continuous at $\xi = 0$, for

$$K(\xi) = \begin{cases} \pi/2, & \xi > 0 \\ 0, & \xi = 0 \\ -\pi/2, & \xi < 0. \end{cases}$$

This implies that the function $|\sin \xi x/x|$ is not dominated (bounded) by an $\mathcal{L}^1(0,\infty)$ function, which clearly follows from the fact that $|\sin x\xi/x|$ is not integrable with respect to x on $(0,\infty)$ (Exercise 1.28). The situation we have here is analogous to the convergence of the Fourier series

$$\sum_{n=1}^\infty \frac{\sin n\xi}{n}$$

to a discontinuous function because its convergence is not uniform. Based on this analogy, the improper integral

$$F(\xi) = \int_a^\infty \varphi(x,\xi)\,dx$$

is said to be *uniformly convergent* on the interval I (see [8]) if, given any $\varepsilon > 0$, there is a number $N > a$ such that

$$b > N \;\Rightarrow\; \left| F(\xi) - \int_a^b \varphi(x,\xi)\,dx \right| = \left| \int_b^\infty \varphi(x,\xi)\,dx \right| < \varepsilon \quad \text{for all } \xi \in I.$$

The number N depends on ε and is independent of ξ.

Corresponding to the Weierstrass M-test for uniform convergence of series, we have the following test for the uniform convergence of improper integrals. We leave the proof as an exercise.

Lemma 6.9 *Let $\varphi : [0, \infty) \times I \to \mathbb{C}$, and suppose that there is a function $g \in \mathcal{L}^1(0, \infty)$ such that $|\varphi(x, \xi)| \le g(x)$ for all ξ in the interval $I = [\alpha, \beta]$. Then the integral $\int_0^\infty \varphi(x, \xi)dx$ is uniformly convergent on I.*

If a function $\varphi(x, \xi)$ satisfies the conditions of Lemma 6.9 and if, in addition, $\varphi(x, \xi)$ is continuous at every $\xi \in I = [\alpha, \beta]$, then the function

$$F(\xi) = \int_a^\infty \varphi(x, \xi)dx$$

is also continuous on I, and satisfies

$$\int_\alpha^\beta F(\xi)d\xi = \int_\alpha^\beta \int_a^\infty \varphi(x, \xi)d\xi dx = \int_a^\infty \int_\alpha^\beta \varphi(x, \xi)d\xi dx. \tag{6.13}$$

This follows from the observation that the uniform convergence

$$\int_a^b \varphi(x, \xi)dx \xrightarrow{u} F(\xi) \text{ as } b \to \infty$$

implies that, for every $\varepsilon > 0$, there is an $N > 0$ (which is independent of ξ) such that, for all $b \ge N$,

$$\left| F(\xi) - \int_a^b \varphi(x, \xi)dx \right| \le \int_b^\infty g(x)dx < \varepsilon$$

$$\Rightarrow \left| \int_\alpha^\beta F(\xi)d\xi - \int_\alpha^\beta \int_0^b \varphi(x, \xi)dx d\xi \right| \le \varepsilon\,(\beta - \alpha).$$

Under the hypothesis of Lemma 6.9, if $\varphi(x, \xi)$ is continuous on $\alpha \le \xi \le \beta$, we can change the order of integration in the double integral

$$\int_a^\infty \int_\alpha^\beta \varphi(x, \xi)d\xi dx = \int_\alpha^\beta \int_0^\infty \varphi(x, \xi)dx d\xi = \int_\alpha^\beta F(\xi)d\xi.$$

This equality remains valid in the limit as $\beta \to \infty$ provided $\int_\alpha^\infty F(\xi)d\xi$ exists.

Pushing the analogy with Fourier series further, we should expect to be able to reconstruct an $\mathcal{L}^1$ function f solely from knowledge of its transform $\hat{f}$, in the same way that a periodic function is determined (according to Theorem 3.9) by its Fourier coefficients. With Eq. (6.6) in mind, we now want to prove

$$f(x) = \frac{1}{2\pi} \int_{-\infty}^\infty \hat{f}(\xi)e^{ix\xi}d\xi,$$

assuming that x is a point of continuity of f. The only problem is that, though continuous and bounded, $\left|\hat{f}\right|$ may not be integrable on $(-\infty, \infty)$, so that the above integral may not converge. Some treatments introduce a damping function, such as $e^{-\varepsilon^2 \xi^2/2}$, into the integrand to force convergence, and then take the limit of the resulting integral as $\varepsilon \to 0$. Here we introduce a cut-off function, already suggested by the Fourier series representation

$$f(x) = \lim_{N \to \infty} \sum_{n=-N}^{N} c_n e^{inx}, \qquad (6.14)$$

which will be further elaborated on in Remark 6.11.

Corresponding to Theorem 3.9 for Fourier series, we have the following fundamental theorem for representing an $\mathcal{L}^1$ function with sufficient smoothness properties by a *Fourier integral*.

Theorem 6.10 *Let f be a piecewise smooth function in $\mathcal{L}^1(\mathbb{R})$. If*

$$\hat{f}(\xi) = \int_{-\infty}^{\infty} f(x) e^{-i\xi x} dx, \quad \xi \in \mathbb{R}, \qquad (6.15)$$

then

$$\lim_{L \to \infty} \frac{1}{2\pi} \int_{-L}^{L} \hat{f}(\xi) e^{ix\xi} d\xi = \frac{1}{2}[f(x^+) + f(x^-)] \ \text{for all } x \in \mathbb{R}. \qquad (6.16)$$

Before attempting to prove this theorem, it is worthwhile to consider some important observations on the meaning and implications of Eq. (6.16).

Remark 6.11

1. The limit

$$\lim_{L \to \infty} \frac{1}{2\pi} \int_{-L}^{L} \hat{f}(\xi) e^{ix\xi} d\xi \qquad (6.17)$$

is the Cauchy principal value of the improper integral

$$\frac{1}{2\pi} \int_{-\infty}^{\infty} \hat{f}(\xi) e^{ix\xi} d\xi = \lim_{\substack{a \to -\infty \\ b \to \infty}} \frac{1}{2\pi} \int_{a}^{b} \hat{f}(\xi) e^{ix\xi} d\xi. \qquad (6.18)$$

This restricted limit (6.17), as is well-known, may exist even when the unrestricted limit (6.18) does not. If $\hat{f}$ lies in $\mathcal{L}^1(\mathbb{R})$ then, of course, the two limits are equal.

2. If f is defined by

$$f(x) = \frac{1}{2}[f(x^+) + f(x^-)] \tag{6.19}$$

at every point of discontinuity x, then (6.16) becomes

$$f(x) = \lim_{L \to \infty} \frac{1}{2\pi} \int_{-L}^{L} \hat{f}(\xi) e^{ix\xi} dx = \mathcal{F}^{-1}(\hat{f}),$$

the inverse Fourier transform of $\hat{f}$, or the Fourier integral of f.
3. When $\hat{f} \in \mathcal{L}^1(\mathbb{R})$ and (6.19) holds, Eqs. (6.15) and (6.16) define the transform pair

$$\hat{f}(\xi) = \mathcal{F}(f)(\xi) = \int_{-\infty}^{\infty} f(x) e^{-i\xi x} dx,$$

$$f(x) = \mathcal{F}^{-1}(\hat{f}) = \frac{1}{2\pi} \int_{-\infty}^{\infty} \hat{f}(\xi) e^{ix\xi} d\xi,$$

a representation which exhibits a high degree of symmetry. Some books adopt the definition

$$\hat{f}(\xi) = \frac{1}{\sqrt{2\pi}} \int_{-\infty}^{\infty} f(x) e^{-ix\xi} dx,$$

from which

$$f(x) = \frac{1}{\sqrt{2\pi}} \int_{-\infty}^{\infty} \hat{f}(\xi) e^{i\xi x} d\xi,$$

and thereby achieve (almost) complete symmetry. Our definition is a natural development of the notation used in Fourier series.
4. If $\hat{f} = 0$ then (6.16) implies $f = 0$. This means that the Fourier transformation $\mathcal{F}$, defined on the piecewise smooth functions in $\mathcal{L}^1(\mathbb{R})$, is injective.

Proof

$$\int_{-L}^{L} \hat{f}(\xi) e^{ix\xi} d\xi = \int_{-L}^{L} \left[\int_{-\infty}^{\infty} f(y) e^{-i\xi y} dy \right] e^{ix\xi} d\xi$$

$$= \int_{-\infty}^{\infty} \int_{-L}^{L} f(y) e^{i\xi(x-y)} d\xi dy,$$

where the change in the order of integration is justified by the fact that the function

$$f(y) e^{i\xi(x-y)}$$

satisfies the conditions of Lemma 6.9. Integrating with respect to ξ, we obtain

$$\int_{-L}^{L} \hat{f}(\xi)e^{ix\xi}\,d\xi = 2\int_{-\infty}^{\infty} \frac{\sin L(x-y)}{x-y} f(y)\,dy$$

$$= 2\int_{-\infty}^{\infty} \frac{\sin L\eta}{\eta} f(x+\eta)\,d\eta. \tag{6.20}$$

Suppose now that δ is an arbitrary positive number. As a function of η, $|f(x+\eta)/\eta|$ is integrable on $(-\infty, -\delta] \cup [\delta, \infty)$, so we can use Lemma 6.5 to conclude that

$$\lim_{L\to\infty} \int_{|\eta|\geq\delta} \frac{\sin L\eta}{\eta} f(x+\eta)\,d\eta = 0.$$

Taking the limits of both sides of Eq. (6.20) as $L \to \infty$, we obtain

$$\lim_{L\to\infty} \int_{-L}^{L} \hat{f}(\xi)e^{ix\xi}\,d\xi = 2\lim_{L\to\infty} \int_{-\delta}^{\delta} \frac{\sin L\eta}{\eta} f(x+\eta)\,d\eta$$

$$= 2\lim_{L\to\infty} \int_{0}^{\delta} \frac{\sin L\eta}{\eta} [f(x+\eta) + f(x-\eta)]\,d\eta.$$

But

$$\lim_{L\to\infty} \int_{0}^{\delta} \frac{\sin L\eta}{\eta} f(x+\eta)\,d\eta = \lim_{L\to\infty} \int_{0}^{\delta} \sin L\eta \frac{f(x+\eta) - f(x^+)}{\eta}\,d\eta$$

$$+ \lim_{L\to\infty} f(x^+) \int_{0}^{\delta} \frac{\sin L\eta}{\eta}\,d\eta.$$

Now Lemma 6.5 implies

$$\lim_{L\to\infty} \int_{0}^{\delta} \sin L\eta \frac{f(x+\eta) - f(x^+)}{\eta}\,d\eta = 0$$

because f is piecewise smooth; and from Lemma 6.8 we conclude that, for any $\delta > 0$,

$$\lim_{L\to\infty} \int_{0}^{\delta} \frac{\sin L\eta}{\eta}\,d\eta = \int_{0}^{\infty} \frac{\sin x}{x}\,dx = \frac{\pi}{2}.$$

Hence

$$\lim_{L\to\infty} \int_{0}^{\delta} \frac{\sin L\eta}{\eta} f(x+\eta)\,d\eta = \frac{\pi}{2} f(x^+).$$

Similarly,

$$\lim_{L \to \infty} \int_0^\delta \frac{\sin L\eta}{\eta} f(x - \eta)d\eta = \frac{\pi}{2} f(x^-).$$

Consequently,

$$\lim_{L \to \infty} \int_{-L}^L \hat{f}(\xi)e^{ix\xi}d\xi = \pi[f(x^+) + f(x^-)],$$

and we obtain the desired equality after dividing by 2π.   □

The similarity between Theorems 3.9 and 6.10 is noteworthy, though it is somewhat obscured by the fact that the trigonometric form of the Fourier series was used in the statement and proof of Theorem 3.9, whereas Theorem 6.10 is expressed in terms of the exponential form of the Fourier transform and integral. The correspondence becomes clearer when the formulas in Theorem 3.9 are cast in the exponential form

$$c_n = \frac{1}{2\pi} \int_{-\pi}^{\pi} f(x)e^{-inx}dx,$$

$$\lim_{N \to \infty} \sum_{n=-N}^{N} c_n e^{inx} = \frac{1}{2}[f(x^+) + f(x^-)].$$

On the other hand, the trigonometric form of Theorem 6.10 is easily derived from Euler's relation $e^{i\theta} = \cos\theta + i\sin\theta$. Suppose $f \in \mathcal{L}^1(\mathbb{R})$ is a real, piecewise smooth function which satisfies

$$f(x) = \frac{1}{2}[f(x^+) + f(x^-)]$$

at each point of discontinuity. Such a function has a Fourier transform

$$\hat{f}(\xi) = \int_{-\infty}^{\infty} f(x)e^{-i\xi x}dx$$

$$= \int_{-\infty}^{\infty} f(x)\cos\xi x dx - i \int_{-\infty}^{\infty} f(x)\sin\xi x \, dx$$

$$= A(\xi) - iB(\xi), \tag{6.21}$$

where

$$A(\xi) = \int_{-\infty}^{\infty} f(x)\cos\xi x \, dx, \tag{6.22}$$

$$B(\xi) = \int_{-\infty}^{\infty} f(x) \sin \xi x \, dx, \quad \xi \in \mathbb{R}, \tag{6.23}$$

are the Fourier cosine and the Fourier sine transforms of f, respectively.

From Theorem 6.10 we can now express $f(x)$ as

$$f(x) = \lim_{L \to \infty} \frac{1}{2\pi} \int_{-L}^{L} [A(\xi) - iB(\xi)]e^{ix\xi} \, d\xi.$$

Since $A(\xi)$ is even and $B(\xi)$ is odd, we have

$$\begin{aligned}
f(x) &= \lim_{L \to \infty} \frac{1}{2\pi} \int_{-L}^{L} [A(\xi) \cos x\xi + B(\xi) \sin x\xi] d\xi \\
&= \lim_{L \to \infty} \frac{1}{\pi} \int_{0}^{L} [A(\xi) \cos x\xi + B(\xi) \sin x\xi] d\xi \\
&= \frac{1}{\pi} \int_{0}^{\infty} [A(\xi) \cos x\xi + B(\xi) \sin x\xi] d\xi, \quad x \in \mathbb{R}. \tag{6.24}
\end{aligned}$$

If the function f is even, then (6.23) implies $B(\xi) = 0$ and the transform pair (6.22) and (6.24) take the form

$$A(\xi) = 2 \int_{0}^{\infty} f(x) \cos \xi x \, dx, \tag{6.25}$$

$$f(x) = \frac{1}{\pi} \int_{0}^{\infty} A(\xi) \cos x\xi \, d\xi. \tag{6.26}$$

If f is odd, then $A(\xi) = 0$ and we obtain

$$B(\xi) = 2 \int_{0}^{\infty} f(x) \sin \xi x \, dx, \tag{6.27}$$

$$f(x) = \frac{1}{\pi} \int_{0}^{\infty} B(x) \sin x\xi \, d\xi. \tag{6.28}$$

The attentive reader will not miss the analogy between these formulas and the corresponding formulas for a_n and b_n in the trigonometric Fourier series.

Example 6.12 Applying Theorem 6.10 to the function $\hat{f}_a(\xi)$ in Example 6.2, we see that

$$\begin{aligned}
\lim_{L \to \infty} \frac{1}{2\pi} \int_{-L}^{L} \frac{2}{\xi} \sin a\xi \, e^{ix\xi} \, d\xi &= \lim_{L \to \infty} \frac{1}{\pi} \int_{-L}^{L} \frac{1}{\xi} \sin a\xi \cos x\xi \, d\xi \\
&= \frac{2}{\pi} \int_{0}^{\infty} \frac{1}{\xi} \sin a\xi \cos x\xi \, d\xi
\end{aligned}$$

since $\sin a\xi/\xi$ is an even function and $\cos\xi x$ is the even (real) part of $e^{i\xi x}$. Therefore, recalling the definition of f_a, we arrive at the following interesting evaluation of the above integral:

$$\frac{2}{\pi}\int_0^\infty \frac{1}{\xi}\sin a\xi \cos x\xi\, d\xi = \begin{cases} 0, & x < -a \\ 1/2, & x = -a \\ 1, & -a < x < a \\ 1/2, & x = a \\ 0, & x > a. \end{cases}$$

Example 6.13 The function

$$f(x) = \begin{cases} \sin x, & |x| < \pi \\ 0, & |x| > \pi \end{cases}$$

is odd. Hence its cosine and sine transforms are, respectively, $A(\xi) = 0$, and

$$\begin{aligned} B(\xi) &= 2\int_0^\infty f(x)\sin\xi x\, dx \\ &= 2\int_0^\pi \sin x \sin\xi x\, dx \\ &= \frac{1}{1-\xi}\sin(1-\xi)\pi - \frac{1}{1+\xi}\sin(1+\xi)\pi \\ &= \frac{2\sin\pi\xi}{1-\xi^2}. \end{aligned}$$

The Fourier integral representation of f is therefore

$$f(x) = \frac{2}{\pi}\int_0^\infty \frac{\sin\pi\xi}{1-\xi^2}\sin x\xi\, d\xi. \tag{6.29}$$

By L'Hôpital's rule,

$$\lim_{\xi\to\pm 1}\frac{\sin\pi\xi}{1-\xi^2}\sin x\xi = \pm\frac{\pi}{2}\sin x,$$

so the integrand $(1-\xi^2)^{-1}\sin\pi\xi\sin x\xi$ is bounded on $0 \le \xi < \infty$, and is dominated by $(1-\xi^2)^{-1}$ as $\xi \to \infty$. The Fourier integral (6.29) therefore converges uniformly on $\mathbb{R}$ to the continuous function f. As shown in Fig. 6.4, both $f(x)$ and $B(x)$ are continuous, but only the Fourier transform is smooth.

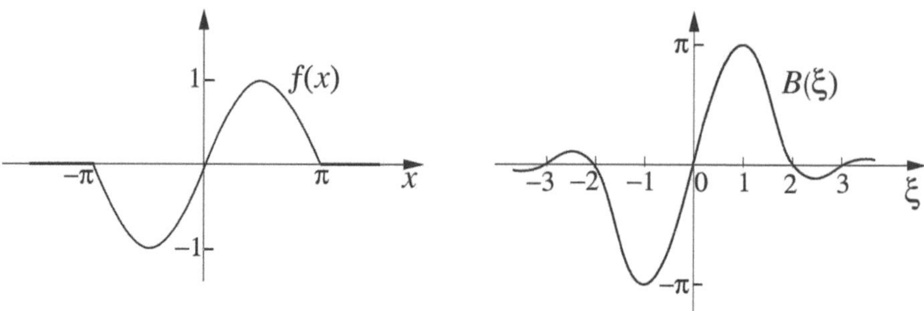

Fig. 6.4 Sine transform of f

Example 6.14 We have already seen in Example 6.3 that the Fourier transform of the even function $e^{-|x|}$ is

$$\mathcal{F}(e^{-|x|})(\xi) = \frac{2}{1+\xi^2},$$

which is an $\mathcal{L}^1$ function on $\mathbb{R}$, hence, using (6.26),

$$e^{-|x|} = \frac{2}{\pi} \int_0^\infty \frac{\cos x\xi}{1+\xi^2} d\xi.$$

At $x = 0$ we obtain the familiar result

$$\int_0^\infty \frac{d\xi}{1+\xi^2} = \frac{\pi}{2}.$$

We conclude this section with a brief look at the Fourier transformation in $\mathcal{L}^2$, the space in which the theory of Fourier series was developed in Chap. 3. In general there is no inclusion relation between $\mathcal{L}^1(\mathbb{R})$ and $\mathcal{L}^2(\mathbb{R})$, but bounded functions in $\mathcal{L}^1(\mathbb{R})$ belong to $\mathcal{L}^2(\mathbb{R})$, since $\int |f|^2 \leq M \int |f|$ whenever $|f| \leq M$. Suppose that both f and g are in $\mathcal{L}^1(\mathbb{R})$, then their transforms $\hat{f}$ and $\hat{g}$ according to Theorem 6.6 are continuous and bounded on $\mathbb{R}$. If we further assume that $\hat{f}, \hat{g} \in \mathcal{L}^1(\mathbb{R})$, then it follows that all four functions are bounded and lie in $\mathcal{L}^2(\mathbb{R})$. In that case we can write

$$2\pi \langle f, g \rangle = 2\pi \int_{-\infty}^\infty f(x)\overline{g(x)}dx$$

$$= \int_{-\infty}^\infty \int_{-\infty}^\infty f(x)\overline{\hat{g}(\xi)e^{ix\xi}}d\xi dx$$

$$= \int_{-\infty}^\infty \int_{-\infty}^\infty f(x)e^{-i\xi x}\overline{\hat{g}(\xi)}dx d\xi$$

$$= \int_{-\infty}^{\infty} \hat{f}(\xi)\overline{\hat{g}(\xi)}d\xi$$

$$= \left\langle \hat{f}, \hat{g} \right\rangle. \tag{6.30}$$

When $g = f$ we obtain a relation between the $\mathcal{L}^2$ norms of f and its Fourier transform,

$$\left\| \hat{f} \right\|^2 = 2\pi \, \|f\|^2 , \tag{6.31}$$

which corresponds to Parseval's relation (1.25). Equations (6.30) and (6.31) together constitute what is known as *Plancherel's Theorem*, which actually holds under weaker conditions on f and g. In fact, this pair of equations is valid whenever f and g lie in $\mathcal{L}^2(\mathbb{R})$. The proof of this more general result is based on the fact that the set of functions $\mathcal{L}^1(\mathbb{R}) \cap \mathcal{L}^2(\mathbb{R})$ is dense in $\mathcal{L}^2(\mathbb{R})$ [9].

Exercises

6.10 Express each of the following functions as a Fourier integral in trigonometric form:

(a) $f(x) = \begin{cases} |\sin x| , & |x| < \pi \\ 0, & |x| > \pi. \end{cases}$

(b) $f(x) = \begin{cases} 1, \ 0 < x < 1 \\ 0, \ x < 0, \ x > 1. \end{cases}$

6.11 Use the result of Exercise 6.10(b) to show that

$$\pi = 2 \int_0^{\infty} \frac{\sin \xi}{\xi} d\xi.$$

6.12 Prove that

$$e^{-\alpha x} = \frac{2}{\pi} \int_0^{\infty} \frac{\xi \sin \xi x}{\xi^2 + \alpha^2} d\xi \quad \text{for all } x > 0, \alpha > 0.$$

Explain why the equality does not hold at $x = 0$.

6.13 Prove that

$$\int_0^{\infty} \frac{\xi^3 \sin x\xi}{\xi^4 + 4} d\xi = \frac{\pi}{2} e^{-x} \cos x \quad \text{for all } x > 0.$$

Is the integral uniformly convergent on $[0, \infty)$?

6.14 Determine the Fourier cosine integral of $e^{-x}\cos x$, $x \geq 0$. Is the representation pointwise?

6.15 Prove that

$$\int_0^\infty \frac{1 - \cos \pi \xi}{\xi} \sin x\xi \, d\xi = \begin{cases} \pi/2, & 0 < x < \pi \\ \pi/4, & x = \pi \\ 0, & x > \pi \end{cases}$$

6.16 Determine the Fourier transform of $f(x) = (1 + x^2)^{-1}$, $x \in \mathbb{R}$.

6.17 Given

$$f(x) = \begin{cases} 1 - |x|, & |x| \leq 1 \\ 0, & |x| > 1, \end{cases}$$

show that

$$\hat{f}(\xi) = \left[\frac{\sin(\xi/2)}{\xi/2} \right]^2 ,$$

then conclude that

$$\pi = \int_{-\infty}^\infty \frac{\sin^2 \xi}{\xi^2} d\xi .$$

6.18 Verify the equality $\left\| \hat{f} \right\|^2 = 2\pi \, \| f \|^2$ in the case where $f(x) = 1$ on $[-1, 1]$, and 0 otherwise..

6.19 Express the relation (6.31) in terms of the cosine and sine transforms A and B.

6.3   Properties and Applications

The following theorem gives the basic properties of the Fourier transformation under differentiation. The formula for the transform of the derivative is particularly important, not only as a tool for solving linear differential equations, but also as a fundamental result on which the existence of solutions to such equations can be based.

Theorem 6.15

(i) If $f \in C^1(\mathbb{R}) \cap L^1(\mathbb{R})$, then

$$\mathcal{F}(f')(\xi) = i\xi \mathcal{F}(f)(\xi), \quad \xi \in \mathbb{R}. \tag{6.32}$$

(ii) If $xf(x) \in \mathcal{L}^1(\mathbb{R})$, then $\mathcal{F}(f)$ is differentiable and its derivative

$$\frac{d}{d\xi}\mathcal{F}(f)(\xi) = \mathcal{F}(-ixf)(\xi), \quad \xi \in \mathbb{R}, \tag{6.33}$$

is continuous on $\mathbb{R}$.

Proof

(i) Since $\left|f'\right|$ is integrable, its Fourier transform $\mathcal{F}(f')$ exists and

$$\mathcal{F}(f')(\xi) = \int_{-\infty}^{\infty} f'(x)e^{-i\xi x}dx.$$

The continuity of f' allows us to write $\int_0^x f'(t)dt = f(x) - f(0)$, and the two limits in the equation

$$\lim_{x \to \pm\infty} f(x) = f(0) + \lim_{x \to \pm\infty} \int_0^x f'(t)dt.$$

exist. But since f is continuous and integrable on $\mathbb{R}$, $\lim_{x \to \pm\infty} f(x) = 0$. Now integration by parts yields

$$\mathcal{F}(f')(\xi) = \left. f(x)e^{-i\xi x}\right|_{-\infty}^{\infty} + i\xi \int_{-\infty}^{\infty} f(x)e^{-i\xi x}dx$$

$$= i\xi\mathcal{F}(f)(\xi).$$

(ii)

$$\frac{\hat{f}(\xi + \Delta\xi) - \hat{f}(\xi)}{\Delta\xi} = \int_{-\infty}^{\infty} f(x)\frac{e^{-i(\xi+\Delta\xi)x} - e^{-i\xi x}}{\Delta\xi}dx.$$

In the limit as $\Delta\xi \to 0$, we obtain

$$\frac{d}{d\xi}\mathcal{F}(f)(\xi) = \lim_{\Delta\xi \to 0} \int_{-\infty}^{\infty} f(x)\frac{e^{-i(\xi+\Delta\xi)x} - e^{-i\xi x}}{\Delta\xi}dx.$$

Since, by hypothesis, the limit

$$\lim_{\Delta\xi \to 0} f(x)\frac{e^{-i(\xi+\Delta\xi)x} - e^{-i\xi x}}{\Delta\xi} = -ixf(x)e^{-i\xi x}$$

is dominated by $|xf(x)| \in \mathcal{L}^1(\mathbb{R})$ for every $\xi \in \mathbb{R}$, we can use Theorem 6.4 to conclude that

$$\frac{d}{d\xi}\mathcal{F}(f)(\xi) = \int_{-\infty}^{\infty} [-ixf(x)]e^{-i\xi x}dx = \mathcal{F}(-ixf)(\xi).$$

$\square$

Using induction, this result can be generalized to higher orders of differentiation.

Corollary 6.16

(i) Suppose $f \in C^n(\mathbb{R}) \cap L^1(\mathbb{R})$ and n is any positive integer, then

$$\mathcal{F}(f^{(n)})(\xi) = (i\xi)^n \mathcal{F}(f)(\xi). \tag{6.34}$$

(ii) If $x^n f(x) \in L^1(\mathbb{R})$, then

$$\frac{d^n}{d\xi^n}\mathcal{F}(f)(\xi) = \mathcal{F}[(-ix)^n f](\xi). \tag{6.35}$$

The integrability of $|x^n f(x)|$ on $\mathbb{R}$ may be viewed as a measure of how fast the function $f(x)$ tends to 0 as $|x| \to \infty$, in the sense that $f(x)$ tends to 0 faster when n is larger. Equation (6.34) therefore implies that, as the order of differentiability (or smoothness) of f increases, so does the rate of decay of $\hat{f}$. Equation (6.35), on the other hand, indicates that functions which decay faster have smoother transforms.

Example 6.17 We can use Theorem 6.15 as a short cut to transform the function $f(x) = e^{-x^2}$, $x \in \mathbb{R}$, of Exercise 6.4. Since f satisfies the conditions of Theorem 6.15, Eq. (6.33) allows us to write

$$\frac{d}{d\xi}\hat{f}(\xi) = -i \int_{-\infty}^{\infty} xe^{-x^2}e^{-i\xi x}dx$$

$$= \frac{i}{2} \int_{-\infty}^{\infty} \frac{d}{dx}\left(e^{-x^2}\right)e^{-i\xi x}dx$$

$$= -\frac{i}{2} \int_{-\infty}^{\infty} e^{-x^2}(-i\xi)e^{-i\xi x}dx = -\frac{\xi}{2}\hat{f}(\xi).$$

The solution of this equation is

$$\hat{f}(\xi) = ce^{-\xi^2/4},$$

where the integration constant is determined by setting $\xi = 0$ and using the equality

$$c = \hat{f}(0) = \int_{-\infty}^{\infty} e^{-x^2}\,dx = \sqrt{\pi}.$$

Thus

$$\mathcal{F}(e^{-x^2})(\xi) = \sqrt{\pi}\,e^{-\xi^2/4}. \tag{6.36}$$

6.3.1 Heat Transfer in an Infinite Bar

Just as Fourier series were used in the construction of solutions to boundary-value problems in bounded space domains, we now show how such solutions can be represented by Fourier integrals. The equations that we have already introduced (Laplace's equation, the heat equation, and the wave equation) all have unique solutions under the boundary conditions imposed, whether the space variable is bounded or not. This is mainly due to the linearity of these equations and that of their boundary conditions. The difference between any two such solutions satisfies a homogeneous differential equation under homogeneous boundary conditions whose only solution is the trivial solution.

Example 6.18 Suppose that an infinite thin bar has an initial temperature distribution along its length given by $f(x)$. We wish to determine the temperature $u(x, t)$ along the bar for all time $t > 0$. To solve this problem by the Fourier transform, we assume that f is piecewise smooth and that $|f|$ is integrable on $(-\infty, \infty)$.

The temperature $u(x, t)$ satisfies the heat equation

$$u_t = k u_{xx}, \quad -\infty < x < \infty,\ t > 0, \tag{6.37}$$

where k is a constant, and the initial condition

$$u(x, 0) = f(x), \quad -\infty < x < \infty, \tag{6.38}$$

specifies the initial temperature distribution along the bar. There are no spacial boundary conditions. As before, we resort to separation of variables by assuming that

$$u(x, t) = v(x)w(t).$$

Substituting into Eq. (6.37), we obtain the equation

$$\frac{v''(x)}{v(x)} = \frac{1}{k}\frac{w'(t)}{w(t)}, \quad -\infty < x < \infty,\ t > 0,$$

which implies that each side must be a constant, say $-\lambda^2$, where $\lambda \in \mathbb{R}$. The resulting pair of equations leads to the following (bounded) solutions

$$v(x) = A(\lambda) \cos \lambda x + B(\lambda) \sin \lambda x,$$

$$w(t) = C(\lambda) e^{-k\lambda^2 t},$$

where A, B, and C are constants of integration which depend on the parameter λ. There are no boundary conditions on the solution, so λ will be a real (rather than a discrete) variable, and $A = A(\lambda)$ and $B = B(\lambda)$ will therefore be functions of λ, and we can set $C(\lambda) = 1$ in the product $v(x)w(t)$. Corresponding to each $\lambda \in \mathbb{R}$, the function

$$u_\lambda(x, t) = [A(\lambda) \cos \lambda x + B(\lambda) \sin \lambda x] e^{-k\lambda^2 t}$$

therefore satisfies Eq. (6.37), $A(\lambda)$ and $B(\lambda)$ being arbitrary functions of $\lambda \in \mathbb{R}$. We do not lose any generality by assuming that $\lambda \geq 0$ since the negative values of λ do not generate additional solutions.

$u_\lambda(x, t)$ cannot be expected to satisfy the initial condition (6.38) for an arbitrary function f, so we assume that the desired solution has the form

$$u(x, t) = \frac{1}{\pi} \int_0^\infty u_\lambda(x, t) d\lambda$$

$$= \frac{1}{\pi} \int_0^\infty [A(\lambda) \cos \lambda x + B(\lambda) \sin \lambda x] e^{-k\lambda^2 t} d\lambda. \tag{6.39}$$

This assumption is suggested by the summation (3.30) over the discrete values of λ, and the factor $1/\pi$ is introduced in order to express u in the form of a Fourier integral. At $t = 0$ we have

$$u(x, 0) = \frac{1}{\pi} \int_0^\infty [A(\lambda) \cos \lambda x + B(\lambda) \sin \lambda x] d\lambda$$

$$= f(x), \quad x \in \mathbb{R}. \tag{6.40}$$

By Theorem 6.10 and Eqs. (6.22)–(6.24), we see that Eq. (6.40) uniquely determines A and B as the cosine and sine transforms, respectively, of f :

$$A(\lambda) = \int_{-\infty}^\infty f(y) \cos \lambda y \, dy$$

$$B(\lambda) = \int_{-\infty}^\infty f(y) \sin \lambda y \, dy.$$

Substituting back into (6.39), and using the even and odd properties of A and B, we arrive at the solution of the heat Eq. (6.37) which satisfies the initial condition (6.38)

$$u(x, t) = \frac{1}{2\pi} \int_{-\infty}^{\infty} \int_{-\infty}^{\infty} f(y)[\cos \lambda y \cos \lambda x + \sin \lambda y \sin \lambda x] e^{-k\lambda^2 t} dy d\lambda$$

$$= \frac{1}{\pi} \int_{0}^{\infty} \int_{-\infty}^{\infty} f(y) \cos(x - y)\lambda \, e^{-k\lambda^2 t} dy d\lambda. \tag{6.41}$$

The solution (6.41) can also be obtained, formally, by first taking the Fourier transform of both sides of the heat equation, as functions of x, and using Corollary 6.16 to obtain

$$\hat{u}_t = k(i\xi)^2 \hat{u}(\xi, t) = -k\xi^2 \hat{u}(\xi, t).$$

The solution of this equation is

$$\hat{u}(\xi, t) = ce^{-k\xi^2 t},$$

where, by the initial condition, $c = \hat{u}(\xi, 0) = \hat{f}(\xi)$. Thus, as a function in ξ,

$$\hat{u}(\xi, t) = \hat{f}(\xi) e^{-k\xi^2 t}$$

is the Fourier transform of $u(x, t)$ for every $t > 0$, and u can therefore be retrieved, according to Theorem 6.10, by the integral

$$u(x, t) = \frac{1}{2\pi} \int_{-\infty}^{\infty} \hat{f}(\xi) e^{-k\xi^2 t} e^{i\xi x} d\xi$$

$$= \frac{1}{2\pi} \int_{-\infty}^{\infty} \left[\int_{-\infty}^{\infty} f(y) e^{-i\xi y} dy \right] e^{-k\xi^2 t} e^{i\xi x} d\xi$$

$$= \frac{1}{2\pi} \int_{-\infty}^{\infty} \int_{-\infty}^{\infty} f(y) e^{i(x-y)\xi} e^{-k\xi^2 t} d\xi dy.$$

Here the fact that $\left| \hat{f}(\xi) \right| e^{-k\xi^2 t}$ is integrable on $-\infty < \xi < \infty$ allows us to replace the Cauchy principal value in (6.16) by the corresponding improper integral, and the change of the order of integration in the last step is justified by the assumption that $|f|$ is integrable. Now, $e^{-k\xi^2 t}$ being an even function of ξ, we have

$$u(x, t) = \frac{1}{\pi} \int_{-\infty}^{\infty} f(y) \int_{0}^{\infty} \cos(x - y)\xi \, e^{-k\xi^2 t} d\xi dy,$$

which coincides with (6.41).

Using the integration formula (see Exercise 6.22)

$$\int_0^\infty e^{-b\xi^2} \cos z\xi \, d\xi = \frac{1}{2}\sqrt{\frac{\pi}{b}} e^{-z^2/4b} \quad \text{for all } z \in \mathbb{R}, \ b > 0, \tag{6.42}$$

we end up with

$$u(x,t) = \frac{1}{2\sqrt{\pi kt}} \int_{-\infty}^\infty f(y) e^{-(y-x)^2/4kt} \, dy$$

$$= \frac{1}{\sqrt{\pi}} \int_{-\infty}^\infty f\left(x + 2\sqrt{kt}\, p\right) e^{-p^2} \, dp.$$

It is straightforward to verify that this last expression for $u(x,t)$ satisfies the heat equation and the initial condition $u(x,0) = f(x)$.

Example 6.19 The corresponding boundary-value problem for a semi-infinite bar, which is insulated at one end, is defined by the system of equations

$$u_t = k u_{xx}, \quad 0 < x < \infty, \ t > 0 \tag{6.43}$$

$$u_x(0,t) = 0, \quad t > 0 \tag{6.44}$$

$$u(x,0) = f(x), \quad 0 < x < \infty. \tag{6.45}$$

The solution of Eq. (6.43) by separation of variables leads to the solutions obtained in Example 6.18,

$$u_\lambda(x,t) = [A(\lambda) \cos \lambda x + B(\lambda) \sin \lambda x] e^{-k\lambda^2 t}, \quad 0 \le \lambda < \infty.$$

For each solution in this set to satisfy the boundary condition at $x = 0$, we must have

$$\frac{\partial u_\lambda}{\partial x}\bigg|_{x=0} = \lambda B(\lambda) e^{-k\lambda^2 t} = 0 \ \text{ for all } t > 0.$$

If $\lambda = 0$ we obtain the constant solution $u_0 = A(0)$, and if $B(\lambda) = 0$ the solution is given by

$$u_\lambda(x,t) = A(\lambda) \cos \lambda x \, e^{-k\lambda^2 t}. \tag{6.46}$$

That means Eq. (6.46), where $0 \le \lambda < \infty$, gives all the solutions of the heat equation which satisfy the boundary condition (6.44).

In order to satisfy the initial condition (6.45), we form the integral of all solutions u_λ,

$$u(x, t) = \frac{1}{\pi} \int_0^\infty u_\lambda(x, t)\, d\lambda = \frac{1}{\pi} \int_0^\infty A(\lambda) \cos x\lambda \; e^{-k\lambda^2 t}\, d\lambda.$$

When $t = 0$,

$$u(x, 0) = f(x) = \frac{1}{\pi} \int_0^\infty A(\lambda) \cos x\lambda \; d\lambda. \tag{6.47}$$

By extending f as an even function into $(-\infty, \infty)$, we see from the representation (6.47) that $A(\lambda)$ is the cosine transform of f. Hence, assuming $f \in \mathcal{L}^1(\mathbb{R})$,

$$A(\lambda) = \int_{-\infty}^\infty f(y) \cos \lambda y \; dy = 2 \int_0^\infty f(y) \cos \lambda y \; dy,$$

and the solution of the boundary-value problem is given by

$$u(x, t) = \frac{2}{\pi} \int_0^\infty \int_0^\infty f(y) \cos \lambda y \cos \lambda x \; e^{-k\lambda^2 t}\, dy d\lambda. \tag{6.48}$$

Using the identity $2 \cos \lambda y \cos \lambda x = \cos \lambda(y - x) + \cos \lambda(y + x)$ and the formula (6.42), this double integral may be reduced to the single integral

$$u(x, t) = \frac{1}{2\sqrt{\pi k t}} \int_0^\infty f(y) \left[e^{-(y-x)^2/4kt} + e^{-(y+x)^2/4kt} \right] dy.$$

To obtain an explicit expression for the solution when

$$f(y) = \begin{cases} 1, & 0 < y < a \\ 0, & y > a, \end{cases}$$

we can use the error function defined in Exercise 5.6,

$$\operatorname{erf}(x) = \frac{2}{\sqrt{\pi}} \int_0^x e^{-p^2}\, dp,$$

to write

$$u(x, t) = \frac{1}{2\sqrt{\pi k t}} \int_0^a \left[e^{-(y-x)^2/4kt} + e^{-(y+x)^2/4kt} \right] dy$$

$$= \frac{1}{\sqrt{\pi}} \int_{-x/2\sqrt{kt}}^{(a-x)/2\sqrt{kt}} e^{-p^2}\, dp + \frac{1}{\sqrt{\pi}} \int_{x/2\sqrt{kt}}^{(a+x)/2\sqrt{kt}} e^{-p^2}\, dp$$

$$= \frac{1}{\sqrt{\pi}} \int_0^{(a-x)/2\sqrt{kt}} e^{-p^2}\,dp + \frac{1}{\sqrt{\pi}} \int_0^{(a+x)/2\sqrt{kt}} e^{-p^2}\,dp$$

$$= \frac{1}{2}\,\mathrm{erf}\left(\frac{a-x}{2\sqrt{kt}}\right) + \frac{1}{2}\,\mathrm{erf}\left(\frac{a+x}{2\sqrt{kt}}\right),$$

as shown in Fig. 6.5. From the properties of the error function it is straightforward to verify that, for all $t > 0$,

$$u(x,t) \to 0 \text{ as } x \to \infty,$$

$$u(x,t) \to \mathrm{erf}\left(a/2\sqrt{kt}\right) \text{ as } x \to 0,$$

and that, for all $x > 0$,

$$u(x,t) \to 0 \text{ as } t \to \infty.$$

As $t \to 0$,

$$u(x,t) \to \begin{cases} 1 & \text{when } 0 \le x < a, \\ 0 & \text{when } x > a. \end{cases}$$

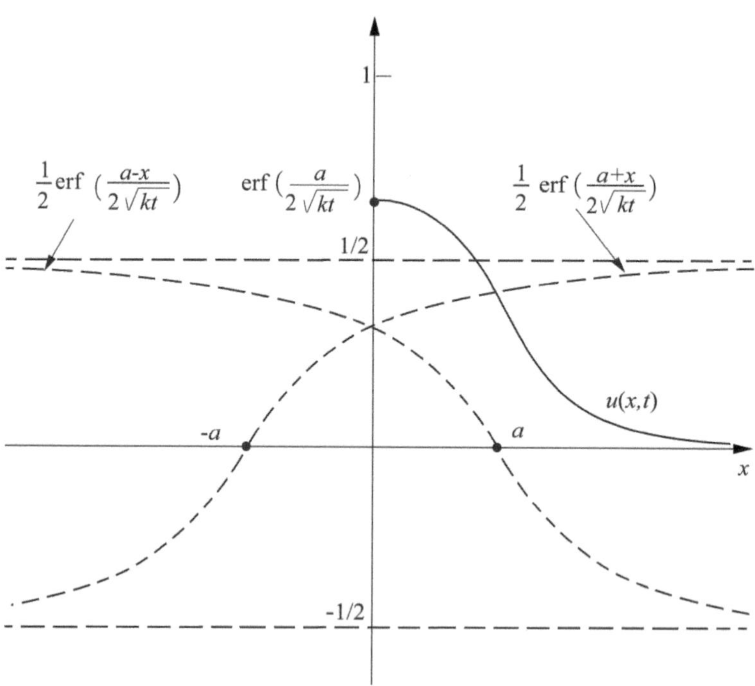

Fig. 6.5 Temperature distribution on a semi-infinite rod

Thus, at any point on the bar, the temperature approaches 0 as $t \to \infty$; and at any instant, the temperature approaches 0 as $x \to \infty$, because the initial heat distribution in the bar (represnted by f) eventually seeps out to ∞. It is also worth noting that $u(a, t) \to 1/2 = [f(a^+) + f(a^-)]/2$ as $t \to 0$, as would be expected.

6.3.2 Non-Homogeneous Equations

The non-homogeneous differential equation $y - y'' = f$ on $\mathbb{R}$ can be solved directly by well-known methods when f is a polynomial or an exponential function, for example, but such methods do not work for more general classes of functions. If f has a Fourier transform, we can use Corollary 6.16 to write

$$\hat{y} + \xi^2 \hat{y} = \hat{f}.$$

The solution of this differential equation is formally given by

$$y(x) = \mathcal{F}^{-1}\left(\hat{f}(\xi)\frac{1}{\xi^2 + 1} \right)(x). \tag{6.49}$$

We know the inverse transforms of both $\hat{f}$ and $(\xi^2 + 1)^{-1}$, but we have no obvious method for inverting the product of these two transforms, or even deciding whether such a product is invertible. Now we show how we can express $\mathcal{F}^{-1}(\hat{f} \cdot \hat{g})$ in terms of f and g under relatively mild restrictions on the functions.

Definition 6.20 Let I be a real interval. A function $f : I \to \mathbb{C}$ is said to be *locally integrable* on I if $|f|$ is integrable on any finite sub-interval of I.

Thus all piecewise continuous functions on I and all functions in $\mathcal{L}^1(I)$ are locally integrable.

Definition 6.21 If the functions $f, g : \mathbb{R} \to \mathbb{C}$ are locally integrable, their *convolution* is the function defined by the integral

$$f * g(x) = \int_{-\infty}^{\infty} f(x - t)g(t)dt$$

for all $x \in \mathbb{R}$ where the integral converges.

By setting $x - t = s$ in the above integral we obtain the commutative relation

$$f * g(x) = \int_{-\infty}^{\infty} f(s)g(x - s)ds = g * f(x).$$

The convolution $f * g$ exists as a function on $\mathbb{R}$ under various conditions on f and g. Here are some examples:

1. If either function is absolutely integrable and the other is bounded. Suppose $f \in \mathcal{L}^1(\mathbb{R})$ and $|g| \le M$. Then

$$|f * g(x)| \le \int_{-\infty}^{\infty} |f(x - t)|\,|g(t)|\,dt$$

$$\le M \int_{-\infty}^{\infty} |f(t)|\,dt < \infty.$$

2. If f and g belong to $\mathcal{L}^2(\mathbb{R})$. This follows directly from the CBS inequality.
3. It can also be shown that if f and g belong to $\mathcal{L}^1(\mathbb{R})$, then $f * g$ belongs to $\mathcal{L}^1(\mathbb{R})$ a.e.

Let $f, g \in \mathcal{L}^1(\mathbb{R})$ and suppose that g is bounded by M, then $f * g(x) \le M \int_{-\infty}^{\infty} |f(x)|\,dx < \infty$, and therefore the convolution integral $f * g(x)$ is uniformly convergent on $\mathbb{R}$. Under these conditions we have the following formula for the Fourier transform of $f * g$.

Theorem 6.22 *If the functions f and g belong to $\mathcal{L}^1(\mathbb{R})$ and one of them is bounded, then the convolution integral $f * g$ is a uniformly convergent on $\mathbb{R}$, and*

$$\mathcal{F}(f * g) = \hat{f} \cdot \hat{g}. \tag{6.50}$$

Proof With $f, g \in \mathcal{L}^1(\mathbb{R})$ we can write

$$\mathcal{F}(f * g)(\xi) = \int_{-\infty}^{\infty} \left[\int_{-\infty}^{\infty} f(x - t)\,g(t)\,dt \right] e^{-i\xi x}dx$$

$$= \int_{-\infty}^{\infty} \left[\int_{-\infty}^{\infty} f(x - t)\,e^{-i\xi(x-t)}g(t)\,e^{-i\xi t}dt \right] dx$$

$$= \int_{-\infty}^{\infty} \int_{-\infty}^{\infty} f(x - t)\,e^{-i\xi(x-t)}e^{-i\xi t}g(t)\,dx dt$$

$$= \hat{f}(\xi) \cdot \hat{g}(\xi),$$

where the change in the order of integration is justified by the absolute integrability of f and g and the uniform convergence of the convolution $f * g$. $\square$

By Theorem 6.6 the transforms $\hat{f}(\xi)$ and $\hat{g}(\xi)$ are both bounded continuous functions on $\mathbb{R}$, which tend to 0 as $|\xi| \to \infty$, so the same applies to the transform of the convolution $\mathcal{F}(f * g)(\xi)$.

Assuming $f \in \mathcal{L}^1(\mathbb{R})$, and using Eq. (6.50) and the result of Example 6.3, we therefore conclude that a particular solution of $y - y'' = f$ is given by

$$
\begin{aligned}
y(x) &= \mathcal{F}^{-1}\left(\hat{f}(\xi)\frac{1}{\xi^2 + 1}\right)(x) \\
&= f * \left(\frac{1}{2}e^{-|\cdot|}\right)(x) \qquad\qquad (6.51) \\
&= \frac{1}{2}\int_{-\infty}^{\infty} f(x - t)e^{-|t|}dt \\
&= \frac{1}{2}\int_{-\infty}^{\infty} f(t)e^{-|x-t|}dt. \qquad\qquad (6.52)
\end{aligned}
$$

We leave it as an exercise to verify, by direct differentiation, that this integral expression satisfies the equation $y - y'' = f$, provided f is continuous.

In this connection, it is worth noting that the kernel function $\frac{1}{2}e^{-|x-t|}$ in the integral representation of y in Eq. (6.52) is none other than Green's function $G(x, t)$ for the SL operator

$$
L = -\frac{d^2}{dx^2} + 1 \qquad\qquad (6.53)
$$

on the interval $(-\infty, \infty)$, subject to the boundary conditions $\lim_{|x| \to \infty} G(x, t) = 0$; for the non-homogeneous equation $Ly = f$ is solved, formally, by

$$
y(x) = L^{-1}f = \int_{-\infty}^{\infty} G(x, t)f(t)dt.
$$

This solution tends to 0 as $x \to \pm\infty$, as it should, being continuous and integrable on $(-\infty, \infty)$. Under such boundary conditions, the solution of the corresponding homogeneous equation $y - y'' = 0$, namely $c_1 e^x + c_2 e^{-x}$, can only be the trivial solution. That implies the solution we have for the non-homogeneous equation is unique.

To solve the same equation $y - y'' = f$ on the semi-infinite interval $[0, \infty)$, we need to impose a boundary condition at $x = 0$, say

$$
y(0) = y_0.
$$

In this case the same procedure above leads to the same particular solution

$$y_p(x) = \frac{1}{2} \int_0^\infty f(t) e^{-|x-t|} dt,$$

but here the (bounded) homogeneous solution

$$y_h(x) = ce^{-x}, \quad x \geq 0$$

is admissible for any constant c. Applying the boundary condition at $x = 0$ to the sum $y = y_p + y_h$ yields

$$y_0 = \frac{1}{2} \int_0^\infty f(t) e^{-t} dt + c,$$

from which c can be determined. The desired solution is therefore

$$y(x) = \frac{1}{2} \int_0^\infty f(t) e^{-|x-t|} dt + \left(y_0 - \frac{1}{2} \int_0^\infty f(t) e^{-t} dt \right) e^{-x}.$$

If $y_0 = 0$, that is, if the boundary condition at $x = 0$ is homogeneous, then

$$y(x) = \frac{1}{2} \int_0^\infty f(t) \left(e^{-|x-t|} - e^{-(x+t)} \right) dt,$$

and, once again, we conclude that Green's function for the same operator (6.53) on the interval $[0, \infty)$ under the homogeneous boundary condition $y(0) = 0$ is now

$$G(x, t) = \frac{1}{2} (e^{-|x-t|} - e^{-(x+t)}).$$

In Chap. 7 we shall see that the Laplace transformation provides a more effective tool for solving non-homogeneous differential equations in $[0, \infty)$, especially where the coefficients are constant.

Exercises

6.20 The *Hermite function* of order n, for any $n \in \mathbb{N}_0$, is defined by

$$\psi_n(x) = e^{-x^2/2} H_n(x), \quad -\infty < x < \infty,$$

where H_n is the Hermite polynomial of order n. Prove that $\hat{\psi}_n$ exists and satisfies

$$\hat{\psi}_n(\xi) = (-i)^n \sqrt{2\pi}\, \psi_n(\xi).$$

Hint: Refer to Example 6.17 and use induction on n

6.21   Solve the integral equation

$$\int_0^\infty u(x)\cos \xi x\, dx = \begin{cases} 1, & 0 < \xi < \pi \\ 0, & \pi < \xi < \infty. \end{cases}$$

6.22   Prove the integration formula (6.42). Hint: Denote the integral by $I(z)$, show that it satisfies the differential equation $2bI'(z) + zI(z) = 0$, then solve this equation.

6.23   Determine the solution of the boundary-value problem

$$u_t = ku_{xx}, \quad 0 < x < \infty,\ t > 0$$

$$u(0, t) = 0, \quad t > 0$$

$$u(x, 0) = f(x), \quad 0 < x < \infty,$$

where f is a piecewise smooth function in $\mathcal{L}^1(0, \infty)$. This is a mathematical model for heat flow in a semi-infinite rod with one end held at zero temperature.

6.24   Solve the wave equation $u_{tt} = c^2 u_{xx}$ on the semi-infinite interval $0 < x < \infty$ under the boundary condition $u(0, t) = 0$ and the initial condition $u(x, 0) = f(x) \in \mathcal{L}^1(0, \infty)$.

6.25   Use the Fourier transform to solve the initial-value problem for the wave equation

$$u_{tt} = c^2 u_{xx}, \quad -\infty < x < \infty,\ t > 0$$

$$u(x, 0) = f(x), \quad -\infty < x < \infty$$

$$u_t(x, 0) = 0, \quad -\infty < x < \infty,$$

where f is assumed to be a piecewise smooth function in $\mathcal{L}^1(\mathbb{R})$. Show that the solution can be expressed as

$$u(x, t) = \frac{1}{2\pi} \int_{-\infty}^\infty \hat{f}(\xi)\cos ct\xi\, e^{ix\xi}\, d\xi,$$

and derive its d'Alembert representation

$$u(x, t) = \frac{1}{2}[f(x + ct) + f(x - ct)].$$

6.26 Use the Fourier transform to solve the wave equation in Exercise 6.25 under the initial conditions $u(x, 0) = f(x)$ and $u_t(x, 0) = g(x)$, where $f, g \in \mathcal{L}^1(\mathbb{R})$ and g is continuous. Derive the d'Alenbert representation of the solution in terms of the functions f and g. Hint: Use Theorem 6.22 and the result of Example 6.12.

6.27 Verify, by direct differentiation, that the integral expression (6.52) solves the equation $y - y'' = f$.

Chapter 7
The Laplace Transformation

If we replace the imaginary variable $i\xi$ in the Fourier transform

$$\hat{f}(\xi) = \int_{-\infty}^{\infty} f(x)e^{-i\xi x}dx$$

by the complex variable $s = \sigma + i\xi$, and set $f(x) = 0$ for all $x < 0$, the function defined by the resulting integral,

$$F(s) = \int_{0}^{\infty} f(x)e^{-sx}dx, \tag{7.1}$$

is called the *Laplace transform* of f. When $\mathrm{Re}\, s = \sigma$ is positive, the improper integral in Eq. (7.1) exists even if $|f|$ is not integrable on $(0, \infty)$, such as when f is a polynomial, and herein lies the advantage of the Laplace transform. Because of the exponential decay of $e^{-\sigma x}$, for $\sigma > 0$, the Laplace transformation is defined on a larger class of functions than $\mathcal{L}^1(0, \infty)$.

7.1 The Laplace Transform

Definition 7.1 We use $\mathcal{E}$ to denote the class of functions $f : [0, \infty) \to \mathbb{C}$ such that f is locally integrable in $[0, \infty)$ and $f(x)e^{-\alpha x}$ remains uniformly bounded as $x \to \infty$ for some real number $\alpha \geq 0$.

Thus a function $f \in \mathcal{E}$ is characterized by two properties: First, that f is allowed to have singular points in $[0, \infty)$ where f is nevertheless locally integrable; and, second, that f has at most exponential growth as $x \to \infty$, i.e., that there is a real non-negative number α such that $f(x)$ is dominated by a constant multiple of the

© The Author(s), under exclusive license to Springer-Verlag London Ltd., part of Springer Nature 2026
M. A. Al-Gwaiz, *Sturm-Liouville Theory and its Applications*, Springer Undergraduate Mathematics Series, https://doi.org/10.1007/978-1-4471-7610-7_7

exponential function $e^{\alpha x}$ for large values of x. More precisely, there are positive constants N and M such that

$$|f(x)| \leq M e^{\alpha x} \quad \text{for all } x \geq N. \tag{7.2}$$

For the sake of convenience, $\mathcal{E}$ will also include any function defined on $\mathbb{R}$ which vanishes on $(-\infty, 0)$ and satisfies Definition 7.1 on $[0, \infty)$. If we now define the *Heaviside function* H on $\mathbb{R}\backslash\{0\}$ (Fig. 7.1) by

$$H(x) = \begin{cases} 1, & x > 0 \\ 0, & x < 0, \end{cases}$$

then the following functions all lie in $\mathcal{E}$:

(i) $H(x)g(x)$ for any bounded locally integrable function g on $\mathbb{R}$, such as $\cos x$ or $\sin x$. Here α, as referred to in Definition 7.1, can be any non-negative number.
(ii) $H(x)p(x)$ for any polynomial p, and α is any positive number.
(iii) $H(x)e^{kx}$, $\alpha \geq k$.
(iv) $H(x)\log x$, α any positive number.
(v) $H(x)x^{\mu}$, $\mu > -1$, α any positive number.

But such functions as $H(x)e^{x^2}$ or $H(x)2^{3^x}$ do not belong to $\mathcal{E}$. The rate of growth of a function f is characterized by the smallest value of α which makes $f(x)e^{-\alpha x}$ bounded as $x \to \infty$. When such a smallest value exists, it is referred to as the exponential order of growth of the function f, such as $\alpha = 0$ in (i) and $\alpha = k$ in (iii). In (ii), (iv), and (v) α can be any positive number.

Definition 7.2 For any $f \in \mathcal{E}$, the *Laplace transform* of f is defined for all $\text{Re}\,s = \sigma > \alpha$ by the integral

$$\mathcal{L}(f)(s) = \int_0^{\infty} f(x)e^{-sx}\,dx, \tag{7.3}$$

which is also denoted by $F(s)$.

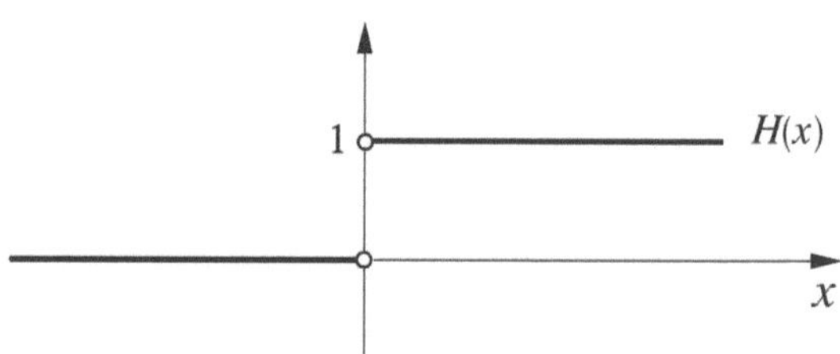

Fig. 7.1 The Heaviside function

Basically, this definition is tailored to ensure that $f(x)e^{-sx}$ lies in $\mathcal{L}^1(0, \infty)$. Using the inequality (7.2),

$$\int_0^\infty \left| f(x)e^{-(\sigma+i\xi)x} \right| dx \leq M \int_0^\infty e^{-(\sigma-\alpha)x} dx < \infty \quad \text{for all } \sigma > \alpha,$$

we see that the integral (7.3) converges for all $\sigma > \alpha$. The convergence is uniform in the half-plane $\sigma \geq \alpha + \varepsilon$ for any $\varepsilon > 0$, where $F(s)$ can be differentiated with respect to s by passing under the integral sign. The resulting integral

$$F'(s) = -\int_0^\infty x f(x)e^{-sx} dx$$

also converges in $\sigma \geq \alpha + \varepsilon$. Consequently the Laplace transform $F = \mathcal{L}(f)$ is an analytic function of the complex variable s in the half-plane $\mathrm{Re}\,s > \alpha$. If $f, g \in \mathcal{E}$, then we clearly have $\mathcal{L}(af + bg) = a\mathcal{L}(f) + b\mathcal{L}(g)$ for any $a, b \in \mathbb{C}$.

Example 7.3 The constant function $f(x) = 1$ on $[0, \infty)$ clearly belongs to $\mathcal{E}$ and its Laplace transform is

$$\mathcal{L}(1)(s) = \int_0^\infty e^{-sx} dx = \frac{1}{s}, \quad s \in \mathbb{C}, \ \mathrm{Re}\,s > 0.$$

Similarly, the function $f : [0, \infty) \to \mathbb{R}$ defined by $f(x) = x^n$, $n \in \mathbb{N}$, also belongs to $\mathcal{E}$. For all $\mathrm{Re}\,s > 0$ we have, using integration by parts,

$$\mathcal{L}(x^n)(s) = \int_0^\infty x^n e^{-sx} dx$$

$$= \frac{n}{s} \int_0^\infty x^{n-1} e^{-sx} dx$$

$$= \cdots$$

$$= \frac{n!}{s^n} \int_0^\infty e^{-sx} dx = \frac{n!}{s^{n+1}}.$$

$$\Rightarrow \quad \mathcal{L}(x^n)(s) = \frac{n!}{s^{n+1}}, \quad n \in \mathbb{N}_0.$$

Since the integrability of $\left| f(x)e^{-sx} \right|$ does not depend on $\mathrm{Im}\,s = \xi$, it is more convenient, for the sake of evaluating the integral (7.3), to set $\mathrm{Im}\,s = 0$. Being analytic on $\mathrm{Re}\,s > \alpha$ in the complex s-plane, the function $F = \mathcal{L}(f)$ is uniquely determined by its values on the real axis.

Example 7.4 For any $\mu > -1$, the Laplace transform

$$\mathcal{L}(x^\mu) = \int_0^\infty x^\mu e^{-sx} dx$$

exists because the singularity of x^μ as $x \to 0^+$ is integrable. To evaluate this indefinite integral, we take s to be real and set $sx = t$. If s is positive,

$$\begin{aligned}
\mathcal{L}(x^\mu) &= \int_0^\infty \left(\frac{t}{s}\right)^\mu e^{-t} \frac{dt}{s} \\
&= \frac{1}{s^{\mu+1}} \int_0^\infty e^{-t} t^\mu dt \\
&= \frac{1}{s^{\mu+1}} \Gamma(\mu + 1), \quad s > 0.
\end{aligned}$$

This result can now be extended analytically into $\mathrm{Re}\, s > 0$, and it clearly generalizes that of Example 7.3.

Example 7.5 For any real number a,

$$\mathcal{L}(e^{ax})(s) = \int_0^\infty e^{-(s-a)x} dx = \frac{1}{s-a}, \quad s > a.$$

Hence,

$$\begin{aligned}
\mathcal{L}(\sinh ax)(s) &= \mathcal{L}\left(\frac{1}{2} e^{ax} - \frac{1}{2} e^{-ax}\right) \\
&= \frac{1}{2} \left(\frac{1}{s-a} - \frac{1}{s+a}\right) \\
&= \frac{a}{s^2 - a^2}, \quad s > a.
\end{aligned}$$

On the other hand,

$$\mathcal{L}(e^{iax})(s) = \int_0^\infty e^{-(s-ia)x} dx = \frac{1}{s - ia}, \quad s > 0.$$

Therefore

$$\mathcal{L}(\sin ax) = \frac{1}{2i} \left(\frac{1}{s-ia} - \frac{1}{s+ia}\right) = \frac{a}{s^2 + a^2},$$

$$\mathcal{L}(\cos ax) = \frac{1}{2} \left(\frac{s}{s-ia} - \frac{s}{s+ia}\right) = \frac{s}{s^2 + a^2}, \quad s > 0.$$

To recover a function f from its Laplace transform $\mathcal{L}(f) = F$, we resort to the Fourier inversion formula (6.16). Suppose $f \in \mathcal{E}$ so that $\left| f(x)e^{-\alpha x} \right|$ is bounded as $x \to \infty$ for some number $\alpha \geq 0$. Choose $\beta > \alpha$ and define the function

$$g(x) = \begin{cases} f(x)e^{-\beta x}, & x \geq 0 \\ 0, & x < 0. \end{cases}$$

g clearly lies in $\mathcal{L}^1(\mathbb{R})$, and its Fourier transform is given by

$$\hat{g}(\xi) = \int_0^\infty f(x)e^{-\beta x}e^{-i\xi x}dx = \mathcal{L}(f)(\beta + i\xi) = F(\beta + i\xi).$$

If we further assume that f is piecewise smooth on $[0, \infty)$, then g will also be piecewise smooth on $\mathbb{R}$, and we can define g to be the average of its left-hand and right-hand limits at each point of discontinuity. By Theorem 6.10,

$$g(x) = f(x)e^{-\beta x}$$

$$= \lim_{L \to \infty} \frac{1}{2\pi} \int_{-L}^{L} F(\beta + i\xi)e^{ix\xi}d\xi, \quad x \geq 0.$$

With β fixed, define the complex variable $s = \beta + i\xi$ so that $ds = id\xi$ and the integral over the interval $(-L, L)$ becomes a contour integral in the complex s-plane over the line segment from $\beta - iL$ to $\beta + iL$. Thus

$$f(x)e^{-\beta x} = \lim_{L \to \infty} \frac{1}{2\pi i} \int_{\beta - iL}^{\beta + iL} F(s)e^{x(s-\beta)}ds,$$

and

$$f(x) = \lim_{L \to \infty} \frac{1}{2\pi i} \int_{\beta - iL}^{\beta + iL} F(s)e^{xs}ds, \quad \beta > \alpha, \tag{7.4}$$

which is the inversion formula for the Laplace transformation $\mathcal{L} : f \mapsto F$. Symbolically, we can write (7.4) as

$$f(x) = \mathcal{L}^{-1}(F)(x),$$

where

$$\mathcal{L}^{-1}(F)(x) = \lim_{L \to \infty} \frac{1}{2\pi i} \int_{\beta - iL}^{\beta + iL} F(s)e^{xs}ds$$

is the inverse Laplace transform of F.

One should keep in mind that (7.4) is a pointwise equality, assuming that $f(x)$ is defined as $\frac{1}{2}\left[f\left(x^+\right) + f\left(x^-\right)\right]$ at any point of discontinuity. It does not apply, for example, to functions which have singularities in $[0, \infty)$, but these are excluded by the assumption that f is piecewise smooth. Furthermore, since $f(0^-) = 0$, we always have

$$\lim_{L \to \infty} \frac{1}{2\pi i} \int_{\beta - iL}^{\beta + iL} F(s)ds = \frac{1}{2}f(0^+). \tag{7.5}$$

The contour integral in (7.4) is not always easy to compute by direct integration; on the contrary, this formula is more often used to evaluate the integral when the function f is known. The same observation applies to (7.5). For example, with reference to the function in Example 7.3, if $\varepsilon > 0$,

$$\lim_{L \to \infty} \frac{1}{2\pi i} \int_{\varepsilon - iL}^{\varepsilon + iL} \frac{1}{s} e^{xs}ds = \begin{cases} 0, & x < 0 \\ 1/2, & x = 0 \\ 1, & x > 0, \end{cases}$$

and the contour integral $\int_{\varepsilon - i\infty}^{\varepsilon + i\infty} s^{-1} e^{xs} ds$ is easily evaluated.

Since $\mathcal{L}^{-1}(0) = 0$, the inverse Laplace transformation is injective on piecewise smooth functions in $\mathcal{E}$. Thus, if $F = \mathcal{L}(f)$ and $G = \mathcal{L}(g)$, where f and g are piecewise smooth functions in $\mathcal{E}$, then $F = G \Rightarrow f = g$.

Exercises

7.1 Determine the Laplace transform $F(s)$ of each of the following functions defined on $(0, \infty)$:

    (a) $f(x) = (ax + b)^2$
    (b) $f(x) = \cosh x$
    (c) $f(x) = \sin^2 x$
    (d) $f(x) = \sin x \cos x$
    (e) $f(x) = \begin{cases} c, & 0 < x < a \\ 0, & a < x < \infty \end{cases}$
    (f) $f(x) = 1/\sqrt{x}$.

7.2 Determine the inverse Laplace transform $f(x)$ for each of the following transform functions:

    (a) $F(s) = \dfrac{s - 1}{(s - 1)^2 + 4}$
    (b) $F(s) = \dfrac{2s - 5}{(s^2 - 9)}$

(c) $F(s) = \dfrac{1}{s(s+1)}$

(d) $F(s) = 1/s^{3/2}$

7.3 If $F(s) = \mathcal{L}(f(x))(s)$ prove that

$$\mathcal{L}(f(ax))(s) = \frac{1}{a}F(s/a)$$

for any $a > 0$. Note the similarity to a corresponding result for the Fourier transform in Exercise 6.7.

7.2   Properties and Applications

As with the Fourier transform, when a function and its derivative have Laplace transforms, the transform functions are related by a formula which can be derived by using integration by parts. We should recall at the outset that a piecewise continuous function is always both locally integrable and locally bounded. Hence all piecewise continuous functions on $[0, \infty)$ of at most exponential growth lie in $\mathcal{E}$.

Theorem 7.6

(i) *If f is a continuous and piecewise smooth function on $[0, \infty)$ such that both $e^{-\alpha x} f(x)$ and $e^{-\alpha x} f'(x)$ are bounded for some $\alpha \geq 0$, then*

$$\mathcal{L}(f')(s) = s\mathcal{L}(f)(s) - f(0^+), \quad \mathrm{Re}\,s > \alpha. \tag{7.6}$$

(ii) *If f is a piecewise continuous function on $[0, \infty)$ and $e^{-\alpha x} f(x)$ is bounded, then*

$$\mathcal{L}\left(\int_0^x f(t)\,dt\right)(s) = \frac{1}{s}\mathcal{L}(f)(s), \quad x > 0, \quad \mathrm{Re}\,s > \alpha. \tag{7.7}$$

Proof

(i) The assumptions on f guarantee the existence of both $\mathcal{L}(f)(s)$ and $\mathcal{L}(f')(s)$ on $\mathrm{Re}\,s > \alpha$. Integrating by parts,

$$\mathcal{L}(f')(s) = \int_0^\infty f'(x)e^{-sx}\,dx$$

$$= e^{-sx} f(x)\big|_0^\infty + s\int_0^\infty f(x)e^{-sx}\,dx$$

$$= s\mathcal{L}(f)(s) - f(0^+), \quad \mathrm{Re}\,s > \alpha.$$

Note that the continuity of f is needed for the second equality to be valid .

(ii) Let

$$g(x) = \int_0^x f(t)dt.$$

Since the function $\left| e^{-\alpha x} f(x) \right|$ is bounded on $[0, \infty)$ by some positive constant, say M, we have

$$|g(x)| \leq M \int_0^x e^{\alpha t} dt \leq \frac{M}{\alpha}(e^{\alpha x} - 1) < \frac{M}{\alpha} e^{\alpha x},$$

hence $|g(x)|$ is bounded on $[0, \infty)$ by $Me^{\alpha x}/\alpha$ if $\alpha > 0$ (and by Mx if $\alpha = 0$), and $\mathcal{L}(g)(s)$ exists for all $\mathrm{Re}\, s > \alpha$. The function g is continuous and its derivative $g' = f$ is piecewise continuous on $[0, \infty)$, so we can use part (i) to conclude that

$$\mathcal{L}(f)(s) = \mathcal{L}(g')(s) = s\mathcal{L}(g) - g(0).$$

With $g(0) = 0$ we arrive at the desired result.

$\square$

By induction on n, we can generalize (7.6) to arrive at the corresponding formula for higher derivatives of f.

Corollary 7.7 *Let $f, f', \cdots, f^{(n-1)}$ be continuous and piecewise smooth functions defined on $[0, \infty)$. If $f, f', \cdots, f^{(n)} \in \mathcal{E}$, then $f^{(n)}$ has a Laplace transform on $\mathrm{Re}\, s > \alpha$, for some $\alpha \geq 0$, given by*

$$\mathcal{L}(f^{(n)})(s) = s^n \mathcal{L}(f)(s) - s^{n-1} f(0^+) - s^{n-2} f'(0^+) - \cdots - f^{(n-1)}(0^+). \qquad (7.8)$$

Example 7.8

(i) Using the result of Example 7.5 and Theorem 7.6,

$$\mathcal{L}(\cos ax)(s) = \mathcal{L}\left(\frac{1}{a}\frac{d}{dx}\sin ax\right)(s)$$

$$= \frac{1}{a}\left(s\frac{a}{s^2 + a^2} - \sin 0\right)$$

$$= \frac{s}{s^2 + a^2}, \quad s > 0.$$

(ii) Given

$$\mathcal{L}(f)(s) = \frac{1}{s(s^2 - 1)}, \quad s > 1,$$

we can apply the formula (7.7) to obtain

$$f(x) = \mathcal{L}^{-1}\left(\frac{1}{s}\frac{1}{s^2-1}\right)(x) = \int_0^x \sinh t\, dt = \cosh x - 1.$$

According to Theorem 7.6, differentiation of f is transformed under $\mathcal{L}$ to multiplication by s, followed by subtraction of $f(0^+)$, whereas integration over $(0, x)$ is transformed to division by s. Conversely, if $f \in \mathcal{E}$ and $\mathcal{L}(f) = F$, then it is easily seen from the definition that multiplication of f by x yields

$$\mathcal{L}(xf)(s) = -\frac{d}{ds}\mathcal{L}(f)(s) = -F'(s).$$

But division by x is a little more tricky because we have to ensure that $f(x)/x$ is integrable in the neibourhood of 0. If $f(x)/x = g(x)$ lies in $\mathcal{E}$ and $\mathcal{L}(g)(s) = G(s)$ then, using the formula above,

$$\mathcal{L}(f)(s) = \mathcal{L}(xg)(s) = -G'(s).$$

Now both transforms F and G are analytic in $\mathrm{Re}\, s > \alpha$, $G' = -F$, and $G(s) \to 0$ as $\mathrm{Re}\, s \to \infty$ (see Exercise 7.12). This uniquely determines G as

$$G(s) = \mathcal{L}(f/x)(s) = \int_s^\infty \mathcal{L}(f)(z)dz, \quad \mathrm{Re}\, s > \alpha, \tag{7.9}$$

where the integral in the half-plane $\mathrm{Re}\, s > \alpha$ is over any contour from s to $+\infty$.

7.2.1 *Applications to Ordinary Differential Equations*

Theorem 7.6 and its corollary make the Laplace transform an effective tool for solving initial-value problems for linear ordinary differential equations, especially when the coefficients in the equation are constant. It reduces the problem to first transforming the differential equation to an algebraic equation, then solving the latter for the transform of the unknown function, and finally inverting the resulting expression of the tranform. Consider, for example, the second order differential equation

$$y'' + ay' + by = f(x), \quad x > 0, \tag{7.10}$$

where a and b are constants, under the initial conditions

$$y(0^+) = y_0, \quad y'(0^+) = y_1. \tag{7.11}$$

Using Corollary 7.7, with $\mathcal{L}(y) = Y$ and $\mathcal{L}(f) = F$, we obtain

$$s^2 Y - s y_0 - y_1 + a(sY - y_0) + bY = F,$$

$$Y(s) = \frac{F(s)}{s^2 + as + b} + \frac{y_0(s + a) + y_1}{s^2 + as + b},$$

and the solution is given by

$$y(x) = \mathcal{L}^{-1}(Y)(x)$$

$$= \mathcal{L}^{-1}\left[\frac{F(s)}{s^2 + as + b}\right](x) + \mathcal{L}^{-1}\left[\frac{y_0(s + a) + y_1}{s^2 + as + b}\right](x)$$

$$= y_p(x) + y_h(x).$$

y_p is a particular solution of the non-homogeneous differential Eq. (7.10), and y_h is the corresponding homogeneous solution which satisfies the initial conditions. This method provides a convenient way for obtaining y_p, especially when the function f is not continuous. The homogeneous part of the solution is simply the inverse transform of a rational function in s, which may be evaluated by using partial fractions, and which vanishes when the initial conditions on y and y' are zero. Thus the non-homogeneous terms in either the differential equation or the boundary condition merely add more terms to the transform function Y, but otherwise pose no extra complication.

Equations (7.10) and (7.11) in fact represent a mathematical model for current flow in an RLC electric circuit. As shown in Fig. 7.2, this consists of a resistance R, an inductance L, and a capacitance C, connected in series and supplied with an input voltage $v(t)$ which depends on time t.

If the current in the circuit is denoted by $y(t)$, then the drop in potential across each of the three circuit elements is given, respectively, by

$$v_R(t) = Ry(t), \quad v_L(t) = Ly'(t), \quad v_C = \frac{1}{C}\int_{t_0}^{t} y(\tau)d\tau.$$

Fig. 7.2 RLC series circuit

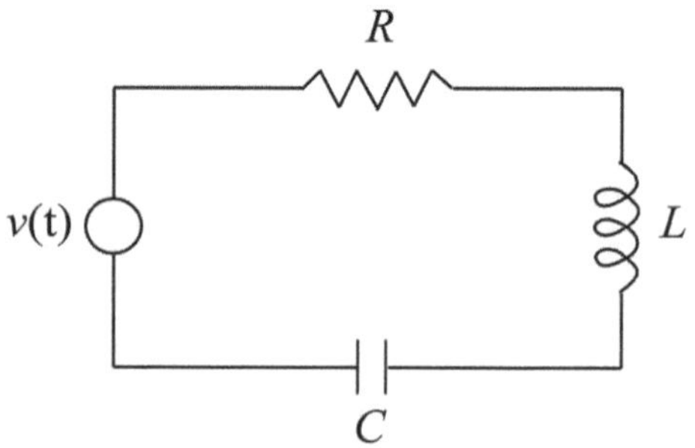

According to Kirchoff's law, the sum of these potential drops is equal to the input voltage,

$$Ry + Ly' + \frac{1}{C}\int_{t_0}^{t} y(\tau)d\tau = v(t).$$

By differentiating with respect to t and dividing by L, we end up with the second order equation

$$y'' + \frac{R}{L}y' + \frac{1}{LC}y = f(t),$$

where the coefficients are constant and $f = v'/L$ is given. If y and y' are specified at $t = 0$, when the circuit switch is closed, then we have a system of equations similar to Eqs. (7.10) and (7.11). The current is determined by solving the system along the lines explained above.

Example 7.9 Solve the initial-value problem

$$y'' + 4y' + 6y = e^{-t}, \quad t > 0,$$

$$y(0) = 0, \quad y'(0) = 0.$$

Solution The Laplace transform of the differential equation is

$$s^2 Y + 4sY + 6Y = \frac{1}{s+1},$$

and the transform function is

$$Y(s) = \frac{1}{(s+1)(s^2+4s+6)} = \frac{1}{(s+1)[(s+2)^2+2]}.$$

To invert this rational function we use partial fractions to write

$$Y(s) = \frac{1}{3}\frac{1}{s+1} - \frac{1}{3}\frac{s+2}{(s+2)^2+2} - \frac{1}{3}\frac{1}{(s+2)^2+2}.$$

By the linearity of the inverse transformation, we therefore have

$$y(t) = \frac{1}{3}e^{-t} - \frac{1}{3}\mathcal{L}^{-1}\left[\frac{s+2}{(s+2)^2+2}\right](t) - \frac{1}{3}\mathcal{L}^{-1}\left[\frac{1}{(s+2)^2+2}\right](t). \tag{7.12}$$

To evaluate the last two expressions, we need to determine the effect of translation in s on the inverse Laplace transform $\mathcal{L}^{-1}$. This is provided by the following translation theorem:

Theorem 7.10 *If $f \in \mathcal{E}$ and $\mathcal{L}(f)(s) = F(s)$ on $\mathrm{Re}\,s > \alpha \geq 0$, then*

$$\mathcal{L}(e^{ax} f)(s) = F(s - a), \quad s - a > \alpha, \tag{7.13}$$

$$\mathcal{L}[H(x - a)f(x - a)](s) = e^{-as} F(s), \quad a \geq 0, \ s > \alpha. \tag{7.14}$$

Proof Equation (7.13) is a direct result of the definition of the Laplace transform,

$$\mathcal{L}(e^{ax} f)(s) = \int_0^\infty f(x)e^{-(s-a)x} dx$$

$$= F(s - a), \quad s - a > \alpha.$$

To prove (7.14), note that

$$e^{-as} F(s) = \int_0^\infty f(x)e^{-s(x+a)} dx$$

$$= \int_a^\infty f(x - a)e^{-sx} dx$$

$$= \int_0^\infty H(x - a)f(x - a)e^{-sx} dx$$

$$= \mathcal{L}[H(x - a)f(x - a)](s).$$

$$\square$$

Going back to Eq. (7.12) in the previous example, since

$$\mathcal{L}^{-1}\left(\frac{s}{s^2 + 2}\right)(t) = \cos \sqrt{2}t$$

$$\mathcal{L}^{-1}\left(\frac{1}{s^2 + 2}\right)(t) = \frac{1}{\sqrt{2}}\mathcal{L}^{-1}\left(\frac{\sqrt{2}}{s^2 + 2}\right) = \frac{1}{\sqrt{2}} \sin \sqrt{2}t,$$

we can apply the formula (7.13) in reverse to obtain

$$\mathcal{L}^{-1}\left[\frac{s + 2}{(s + 2)^2 + 2}\right](t) = e^{-2t} \cos \sqrt{2}t$$

$$\mathcal{L}^{-1}\left[\frac{1}{(s + 2)^2 + 2}\right](t) = \frac{1}{\sqrt{2}}e^{-2t} \sin \sqrt{2}t.$$

The solution (7.12) is therefore

$$y(t) = \frac{1}{3}e^{-t} - \left(\frac{1}{3} \cos \sqrt{2}t + \frac{1}{3\sqrt{2}} \sin \sqrt{2}t\right) e^{-2t}.$$

Remark 7.11 In Example 7.9 the second order equation has to have two initial conditions at $t = 0$ to determine its solution, while similar equations which were discussed in Chap. 6 required only one condition, namely $y(0) = y_0$. The reason is that, in using the Fourier transformation, we are tacitly assuming that y is an $\mathcal{L}^1$ function and, being continuous, tends to 0 as $x \to \infty$. This constitutes a second boundary condition. The Laplace transformation only requires that $y \in \mathcal{E}$, which allows for exponential growth.

In the above example we could just as well have used the conventional method for solving differential equations with constant coefficients. The particular solution $e^{-t}/3$ can easily be obtained by assuming that it is a constant multiple of e^{-t} and substituting into the differential equation to determine its coefficient. But in the following example, where the non-homogeneous term is not continuous, there is a clear advantage to using the Laplace transform.

Example 7.12 Solve the initial-value problem

$$y' + 3y = f(t), \quad t > 0, \tag{7.15}$$

$$y(0) = 1,$$

where

$$f(t) = \begin{cases} 0, & t < 0 \\ t, & 0 < t < 1 \\ 0, & t > 1. \end{cases}$$

Here $y(t)$ represents the current in an RL circuit where the input voltage $f(t) = t$ is switched on at $t = 0$, and then switched off at $t = 1$ (Fig. 7.3).

Solution By expressing f in the form $f(t) = t[H(t) - H(t - 1)]$, we can use the formula $\mathcal{L}(xf) = -F'(s)$ and Theorem 7.10 to write

$$\mathcal{L}(f)(s) = -\frac{d}{ds}\mathcal{L}[(H(t) - H(t - 1))](s)$$

$$= -\frac{d}{ds}\left[\frac{1}{s} - \frac{e^{-s}}{s}\right] = \frac{1}{s^2} - \frac{e^{-s}}{s^2} - \frac{e^{-s}}{s}.$$

Fig. 7.3 Voltage supply in RL circuit

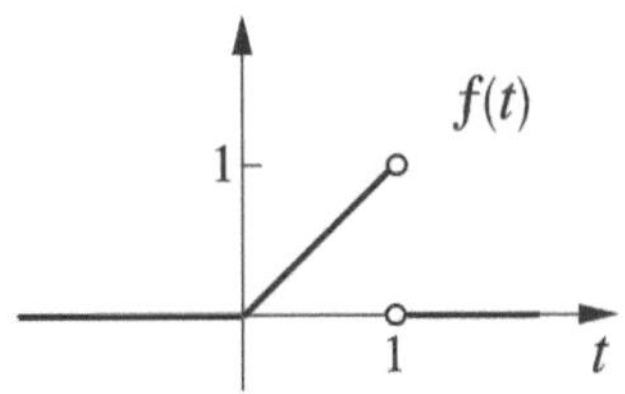

The transformed Eq. (7.15) therefore has the form

$$(s+3)Y(s) - 1 = \frac{1}{s^2} - \frac{e^{-s}}{s^2} - \frac{e^{-s}}{s}$$

$$Y(s) = \frac{1}{s+3} + \frac{1}{s^2}\frac{1}{s+3} - e^{-s}\frac{1}{s^2}\frac{1}{s+3} - e^{-s}\frac{1}{s}\frac{1}{s+3}.$$

With

$$\mathcal{L}^{-1}\left(\frac{1}{s+3}\right) = e^{-3t}H(t),$$

we use (7.7) to write

$$\mathcal{L}^{-1}\left(\frac{1}{s}\frac{1}{s+3}\right)(t) = \int_0^t e^{-3\tau}d\tau = \frac{1}{3}(1 - e^{-3t})H(t)$$

$$\mathcal{L}^{-1}\left(\frac{1}{s^2}\frac{1}{s+3}\right)(t) = \int_0^t \frac{1}{3}(1 - e^{-3\tau})d\tau = \frac{1}{3}\left(t + \frac{1}{3}e^{-3t} - \frac{1}{3}\right)H(t).$$

Now (7.14) implies

$$\mathcal{L}^{-1}\left(e^{-s}\frac{1}{s}\frac{1}{s+3}\right)(t) = \frac{1}{3}[1 - e^{-3(t-1)}]H(t-1)$$

$$\mathcal{L}^{-1}\left(e^{-s}\frac{1}{s^2}\frac{1}{s+3}\right)(t) = \frac{1}{3}\left[(t-1) + \frac{1}{3}e^{-3(t-1)} - \frac{1}{3}\right]H(t-1),$$

hence

$$y(t) = \frac{1}{3}\left(t + \frac{10}{3}e^{-3t} - \frac{1}{3}\right)H(t) - \frac{1}{3}\left(t - \frac{2}{3}e^{-3(t-1)} - \frac{1}{3}\right)H(t-1)$$

$$= \begin{cases} \dfrac{1}{3}\left(t - \dfrac{1}{3}\right) + \dfrac{10}{9}e^{-3t}, & 0 < t \le 1 \\[2mm] \dfrac{10 + 2e^3}{9}e^{-3t}, & t > 1. \end{cases}$$

Although f is discontinuous at $x = 1$, the solution y is continuous on $(0, \infty)$. This is to be expected, since the inductance L, represented by the presence of y' in the differential equation, does not allow sudden changes in the current in an electric circuit by introducing a damping factor into the solution, in this case e^{-3t}.

The method used to arrive at the solutions in Examples 7.8, 7.9, and 7.12 forms part of a collection of techniques based on the properties of the Laplace transformation, known as the *operational calculus*, which was developed by the English physicist *Oliver Heaviside* (1850–1925) to solve linear differential equations of electric circuits. This method is not limited to equations with constant coefficients, as the next example shows.

Example 7.13 Under the Laplace transformation, Laguerre's equation

$$xy'' + (1 - x)y' + ny = 0, \quad x > 0, \ n \in \mathbb{N}_0,$$

becomes

$$-\frac{d}{ds}[s^2 Y - sy(0) - y'(0)] + sY - y(0) + \frac{d}{ds}[sY - y(0)] + nY = 0$$

$$(s - s^2)Y' + (n + 1 - s)Y = 0$$

$$\frac{Y'}{Y} = \frac{n + 1 - s}{s(s - 1)} = \frac{n}{s - 1} - \frac{n + 1}{s}.$$

Integrating the last equation, we obtain

$$Y(s) = c\frac{(s - 1)^n}{s^{n+1}},$$

where c is the integration constant. Taking the inverse transform,

$$y(x) = c\mathcal{L}^{-1}\left[\frac{(s - 1)^n}{s^{n+1}}\right](x)$$

$$= ce^x\mathcal{L}^{-1}\left[\frac{s^n}{(s + 1)^{n+1}}\right](x)$$

$$= \frac{c}{n!}e^x\frac{d^n}{dx^n}(e^{-x}x^n).$$

Laguerre's polynomial L_n is obtained by setting $c = 1$.

7.2.2 The Telegraph Equation

The Laplace transform can also be used to solve boundary-value problems for partial differential equations, especially where the time variable t is involved; for then the transform can naturally be applied with respect to t over the semi-infinite interval $[0, \infty)$. Consider the linear equation

$$u_{xx} = Au_{tt} + Bu_t + Cu, \quad 0 < t < \infty, \tag{7.16}$$

where $A, B,$ and C are non-negative constants (which do not all vanish). This equation, known as the *telegraph equation*, describes an electromagnetic signal $u(x, t)$, such as an electric current or voltage, traveling along a transmission line. The constants A, B, C are determined by the distributed inductance, resistance and capacitance (per unit length) along the line (see [7], vol 2). If the transmission line extends over $-\infty < x < \infty$, two initial conditions (at $t = 0$) on u and u_t are sufficient to specify u. But if it extends over $0 \leq x < \infty$, then we need to specify u at $x = 0$ as well, in which case the boundary conditions take the form

$$u(0, t) = f(t), \quad t \geq 0,$$

$$u(x, 0) = g(x), \quad x \geq 0,$$

$$u_t(x, 0) = h(x), \quad x \geq 0.$$

Here $f(t)$ is the input signal which is transmitted, and the two initial conditions are needed because Eq. (7.16) is of second order in t (if $A = 0$ we would not need the condition on u_t). We can attempt to solve Eq. (7.16) by separation of variables, using the Fourier integral as we did previously, but the fact that all the boundary conditions are non-homogeneous makes the procedure more difficult. It also places unnecessary restrictions on the behaviour of u and f as $t \to \infty$.

We can simplify things by solving Eq. (7.16) under two sets of boundary conditions: first with $f = 0$, and then with $g = h = 0$. Since (7.16) is linear and homogeneous, the sum of these two solutions is also a solution of the equation, and it satisfies the three non-homogeneous boundary conditions. In both cases, the Laplace transform can be used to construct the solution. Let us take the second case, where the boundary conditions are

$$u(0, t) = f(t), \quad t \geq 0,$$

$$u(x, 0) = u_t(x, 0) = 0, \quad x \geq 0.$$

Assuming that $f, u, u_x, u_{xx}, u_t,$ and u_{tt} all lie in $\mathcal{E}$ as functions of t, and that u and its first partial derivatives are piecewise smooth functions of t, we can apply the Laplace transformation with respect to t to Eq. (7.16) to obtain

$$U_{xx} = (As^2 + Bs + C)U, \tag{7.17}$$

where

$$U(x, s) = \int_0^\infty u(x, t)e^{-st}dt \tag{7.18}$$

and $\mathrm{Re}\, s > \alpha \geq 0$. If, for example, the signal u is bounded we may take $\alpha = 0$. Note that differentiation with respect to x may be taken inside the integral (7.18) since the latter converges uniformly on $\mathrm{Re}\, s \geq \alpha + \varepsilon$ for any $\varepsilon > 0$. For every fixed value of the parameter s, Eq. (7.17) has the general solution

$$U(x, s) = c_1(s)e^{\lambda(s)x} + c_2(s)e^{-\lambda(s)x},$$

where $c_1 + c_2 = U(0, s) = \mathcal{L}(f)(s) = F(s)$ and $\lambda(s) = \sqrt{As^2 + Bs + C}$. If we take the principal branch of the square root, where $\mathrm{Re}\lambda(s) > 0$, the solution $e^{\lambda(s)x}$ becomes unbounded in the right half of the s-plane and must therefore be dropped (by setting $c_1 = 0$). The resulting transform function

$$U(x, s) = F(s)e^{-\lambda(s)x}$$

can, in principle, be inverted to yield the desired solution $u(x, t) = \mathcal{L}^{-1}(U)(x, t)$, though the computations involved may be quite tedious. Nevertheless, there are a number of significant special cases of Eq. (7.16) where the solution can be obtained explicitly:

(i) If $A = 1/c^2$ and $B = C = 0$, (7.16) reduces to the familiar wave equation

$$u_{tt} = c^2 u_{xx}.$$

Here the transform function

$$U(x, s) = e^{-sx/c} F(s)$$

is easily inverted, using formula (7.14), to

$$u(x, t) = H(t - x/c)f(t - x/c).$$

This expression indicates that there is no signal at a point x until $t = x/c$. Thus the signal $f(t)$ moves down the line without distortion or damping and with velocity c. This is to be expected, since the damping factor Bu_t is absent and, consequently, no energy is dissipated in the transmission.

(ii) If $As^2 + Bs + C = (s + b)^2/c^2$ is a perfect square, then

$$U(x, s) = e^{-(s+b)x/c} F(s)$$

and the solution is

$$u(x, t) = H(t - x/c)e^{-bx/c}f(t - x/c).$$

Here the signal $f(t)$ again moves down the line with velocity c, undistorted, but attenuated by the damping factor $e^{-bx/c}$. A transmission line therefore becomes distortionless if it is designed so that the distributed resistance, inductance, and capacitance are such that the relation $B^2 = 4AC$ holds. The interested reader may refer to [10] on this point and on some other interesting historical background to the telegraph equation.

(iii) If $A = C = 0$ and $B = 1/k$, where k is a positive constant, we obtain the heat equation

$$u_t = ku_{xx}.$$

In this case the Laplace transform of u is

$$U(x, s) = e^{-x\sqrt{s/k}}F(s), \tag{7.19}$$

which is not readily invertible. To determine $\mathcal{L}^{-1}(U)$ we need another rule which allows us to invert the product of two transforms. By analogy with the Fourier transformation we should expect this to involve the convolution of the inverse transforms.

If the functions f and g are locally integrable on $[0, \infty)$, then their convolution

$$f * g(x) = \int_0^x f(x - t)g(t)dt = \int_0^x f(t)g(x - t)dt = g * f(x)$$

is well defined and locally integrable on $[0, \infty)$. For if $[a, b]$ is any finite interval in $[0, \infty)$, then

$$\int_a^b |f * g(x)|\, dx \le \int_a^b \int_0^x |f(x - t)|\,|g(t)|\, dtdx$$

$$= \int_a^b \int_0^b H(x - t)\,|f(x - t)|\,|g(t)|\, dtdx$$

$$= \int_0^b \left[\int_a^b H(x - t)\,|f(x - t)|\, dx\right]|g(t)|\, dt$$

$$\le \int_0^b \left[\int_{a-t}^{b-t} H(y)\,|f(y)|\, dy\right]|g(t)|\, dt.$$

Because f is locally integrable and $0 \le t \le b$, the integral of $H\,|f|$ over $[a-t, b-t]$ is uniformly bounded by a constant. The function g is also integrable on $[0, b]$, therefore $|f * g|$ is integrable on $[a, b]$.

If $f(x)$ and $g(x)$ are dominated as $x \to \infty$ by $e^{\alpha x}$, then $|f(x - t)\,g(t)|$ is dominated by $e^{\alpha(x-t)}e^{\alpha t} = e^{\alpha x}$. Consequently, if f and g belong to $\mathcal{E}$, then so does their convolution $f * g$, and its Laplace transform is given by

$$\mathcal{L}(f * g)(s) = \int_0^\infty e^{-sx} \int_0^x f(t)g(x - t)dtdx$$

$$= \int_0^\infty \int_0^\infty H(x - t) f(t) g(x - t) e^{-sx} \, dt \, dx$$

$$= \int_0^\infty f(t) \int_0^\infty g(y) e^{-s(t+y)} \, dy \, dt$$

$$= \mathcal{L}(f)(s) \cdot \mathcal{L}(g)(s).$$

In the third equality, the order of integration is reversed, and this is justified by the uniform convergence of the convolution integral on $\mathrm{Re}\, s \geq \alpha + \varepsilon$ for any positive ε. Thus we have proved the following convolution theorem which corresponds to Theorem 6.22 for the Fourier transformation:

Theorem 7.14 *Let $f, g \in \mathcal{E}$. If $\mathcal{L}(f)(s) = F(s)$ and $\mathcal{L}(g)(s) = G(s)$, then*

$$\mathcal{L}(f * g)(s) = F(s)G(s).$$

Now we can go back to Eq. (7.19) to conclude that

$$u(x, t) = \mathcal{L}^{-1}\left(e^{-x\sqrt{s/k}} F(s)\right)(t)$$

$$= f * \mathcal{L}^{-1}\left(e^{-x\sqrt{s/k}}\right)(t).$$

The function $\mathcal{L}^{-1}(e^{-x\sqrt{s/k}})(t)$ can be evaluated by using the inversion formula (7.4), which requires some manipulations of contour integrals (see Exercise 7.21), or it may be looked up in a table of Laplace transforms. In either case

$$\mathcal{L}^{-1}\left(e^{-x\sqrt{s/k}}\right)(t) = \frac{x}{\sqrt{4\pi k t^3}} e^{-x^2/4kt}, \tag{7.20}$$

hence

$$u(x, t) = \frac{x}{\sqrt{4\pi k}} \int_0^t f(t - \tau) \tau^{-3/2} e^{-x^2/4k\tau} \, d\tau. \tag{7.21}$$

Here the solution u differs considerably from that in the first two cases. It tends to 0 as $x \to \infty$ at any time t and also as $t \to \infty$ at any point x, and the signal f is distorted as it spreads along the line with no fixed velocity. By dropping the first term in (7.16), the equation is radically changed from a hyperbolic equation, with wave-like solutions, to a parabolic equation whose solutions travel in a diffusive manner, in much the same way that heat or gas spreads out.

Exercises

7.4 Sketch each of the following functions and determine its Laplace transform:

(a) $(x-1)H(x-1)$
(b) $(x-1)^2 H(x-1)$
(c) $x^2[H(x-1) - H(x-3)]$
(d) $H(x-\pi/2)\cos x$
(e) $(1 - e^{-x})[H(x) - H(x-1)]$.

7.5 Use the Heaviside function to represent the function

$$f(x) = \begin{cases} x, & 0 < x < 1 \\ e^{1-x}, & x > 1 \end{cases}$$

by a single expression. Sketch the function and determine its Laplace transform.

7.6 Determine the inverse Laplace transform of each of the following functions:

(a) $\dfrac{e^{-6s}}{s^3}$

(b) $\dfrac{e^{-s}}{s^2 + 2s + 2}$

(c) $\dfrac{1}{s}(e^{-3s} + e^{-s})$

(d) $\dfrac{1}{s-1}(e^{-3s} + e^{-s})$

(e) $\dfrac{1 + e^{-\pi s}}{s^2 + 9}$.

7.7 If f is not assumed to be continuous in Theorem 7.6(i), how would Eq. (7.6) look like?

7.8 Solve the following initial-value problems:

(a) $y'' + 4y' + 5y = 0, \quad y(0) = 1, \quad y'(0) = 1$
(b) $9y'' - 6y' + y = 0, \quad y(0) = 3, \quad y'(0) = 1$
(c) $y'' + 2y' + 5y = 3e^{-x}\sin x, \quad y(0) = 0, \quad y'(0) = 3$
(d) $y'' + 2y' - 8y = -256x^3, \quad y(0) = 15, \quad y'(0) = 36$
(e) $y'' - 3y' + 2y = H(x-1), \quad y(0) = 0, \quad y'(0) = 1$
(f) $y' + 2y = H(x) - H(x-1), \quad y(0) = 0$.

7.9 Invert each of the following transforms:

(a) $\dfrac{s}{(s^2 + 9)^2}$

(b) $\log\left(\dfrac{s}{s-1}\right)$

(c) $\log\left(\dfrac{s+a}{s+b}\right)$

(d) $\operatorname{arccot}(s+1)$.

7.10 The function $\operatorname{Si} : \mathbb{R} \to \mathbb{R}$, defined by the improper integral

$$\operatorname{Si}(x) = \int_0^x \frac{\sin t}{t}\,dt,$$

is called the *sine integral*. Prove that

$$\mathcal{L}[\operatorname{Si}(x)](s) = \frac{1}{s}\arctan\left(\frac{1}{s}\right).$$

7.11

(a) Let $f \in \mathcal{E}$ be a periodic function on $[0, \infty)$ of period $p > 0$. Show that

$$\mathcal{L}(f)(s) = \frac{1}{1 - e^{-ps}}\int_0^p f(x)e^{-sx}dx, \quad \operatorname{Re}s > 0.$$

(b) Use this to compute $\mathcal{L}(f)$, where $f(x) = x$ on $(0, 1)$, $f(x+1) = f(x)$ for all $x > 0$, and $f(x) = 0$ for all $x < 0$.

7.12 Prove that $\mathcal{L}(f)(s) \to 0$ as $\operatorname{Re}s \to \infty$ for any $f \in \mathcal{E}$.

7.13 Use the Laplace transformation to solve the integral equation $\int_0^x (x - t)^3 y(t)dt = f(x)$, and state the conditions on f which insure that the method works.

7.14 If $f(x) = e^x$ and $g(x) = 1/\sqrt{\pi x}$ prove that

$$f * g(x) = e^x \operatorname{erf}(\sqrt{x}).$$

Use this to evaluate $\mathcal{L}[e^x \operatorname{erf}(\sqrt{x})]$ and $\mathcal{L}[\operatorname{erf}(\sqrt{x})]$.

7.15 Determine $\mathcal{L}([x])$, where $[x]$ is the integer part of the non-negative real number x, i.e.,

$$[x] = n \quad \text{for all } x \in [n, n+1), \ n \in \mathbb{N}_0.$$

7.16 Given two locally integrable functions $f, g : [0, \infty) \to \mathbb{C}$, show that

(a) $f * g$ is continuous if either f or g is continuous.

(b) $f * g$ is smooth if either f or g is smooth and the other continuous.

7.17 Use the Laplace transformation to obtain the solution of the wave equation

$$u_{tt} = c^2 u_{xx}, \quad x > 0, \ t > 0,$$

subject to the boundary condition

$$u(0, t) = \cos^2 t, \quad t \geq 0,$$

and the initial conditions

$$u(x, 0) = u_t(x, 0) = 0, \quad x > 0.$$

7.18 Solve the boundary-value problem

$$u_t - au = ku_{xx}, \quad x > 0, \ t > 0,$$
$$u(0, t) = f(t), \quad t \geq 0,$$
$$u(x, 0) = 0, \quad x \geq 0.$$

7.19 Solve the boundary-value problem for the telegraph equation

$$u_{tt} + 2u_t + u = c^2 u_{xx}, \quad x > 0, \ t > 0,$$
$$u(0, t) = \sin t, \quad t \geq 0,$$
$$u(x, 0) = u_t(x, 0) = 0, \quad x \geq 0.$$

7.20 In Eq. (7.21), prove that $u(x, t) \to f(t)$ as $x \to 0$.

7.21 Use contour integration in the complex s-plane to prove that

$$\mathcal{L}^{-1}(e^{-a\sqrt{s}}/\sqrt{s})(x) = \frac{1}{\sqrt{\pi x}} e^{-a^2/4x},$$

where $a > 0$, then conclude that

$$\mathcal{L}^{-1}(e^{-a\sqrt{s}}) = \frac{a}{2\sqrt{\pi x^3}} e^{-a^2/4x}.$$

Appendix A
Lebesgue Integration

Before introducing the Lebesgue integral, it would be instructive to recall how the Riemann integral was defined. For a real bounded function f defined on a bounded interval $[a, b]$, we first divide the interval into n sub-intervals by a partition P of points $\{x_i\}$ such that

$$a = x_0 < x_1 < x_2 < \cdots < x_n = b,$$

to form the *Riemann sum*

$$\sum_{i=0}^{n-1} f\left(\xi_i\right) (x_{i+1} - x_i),$$

where ξ_i is any point in $[x_i, x_{i+1}]$. Let

$$M_i = \sup\{f(x) : x \in [x_i, x_{i+1}]\},$$
$$m_i = \inf\{f(x) : x \in [x_i, x_{i+1}]\},$$

and define the *Riemann upper sum* and the *Riemann lower sum* of f associated with the partition P, respectively, as

$$U(f, P) = \sum_{i=0}^{n-1} M_i (x_{i+1} - x_i),$$

$$L(f, P) = \sum_{i=0}^{n-1} m_i (x_{i+1} - x_i).$$

© The Author(s), under exclusive license to Springer-Verlag London Ltd., part of Springer Nature 2026
M. A. Al-Gwaiz, *Sturm-Liouville Theory and its Applications*, Springer Undergraduate Mathematics Series, https://doi.org/10.1007/978-1-4471-7610-7

For a positive function f over $[a, b]$ the intuitive notion of *area A* under the curve $y = f(x)$ over this interval clearly satisfies $L(f, P) \leq A \leq U(f, P)$ for all partitions of $[a, b]$.

Now let $\|P\| = \max\{x_{i+1} - x_i : i = 0, 1, 2, \cdots, n-1\}$. If the partition P is refined by adding more points, the difference between the resulting upper and lower sums can only decrease. If they approach a common limit as $n \to \infty$ and $\|P\| \to 0$, that limit is defined to be the *Riemann integral* of f on $[a, b]$, denoted by $\int_a^b f(x)\,dx$, otherwise the integral does not exist. Here the intuitive notion of "area" provides a geometric interpretation of the integral of the function f just as the "slope" of its curve represented a geometric interpretation of its derivative in differential calculus. Alternatively we can also think of the derivative and the integral as providing analytical definitions of these intuitive concepts.

Any function on a finite interval is Riemann integrable if it is continuous except possibly at a finite number of points $\{x_i\}$, where the one-sided limits $f\left(x_i^{\pm}\right)$ exist, that is, where the discontinuity is a simple "jump", regardless of how it is defined at the point of discontinuity. But not every bounded function on a bounded interval is Riemann integrable, as in the case of the Dirichlet function $f : [0, 1] \to \mathbb{R}$,

$$f(x) = \begin{cases} 1, & x \in \mathbb{Q} \cap [0, 1] \\ 0, & x \notin \mathbb{Q} \cap [0, 1], \end{cases} \tag{A.1}$$

where $\mathbb{Q}$ is the set of rational numbers. Here, for any partition of $[0, 1]$, $m_i = 0$ and $M_i = 1$ for all i, therefore the upper sum is 1 whereas the lower sum is 0 for any partition, so their difference does not converge to 0 and therefore the function is not Riemann integrable. The set of Riemann integrable functions on $\Omega \subseteq \mathbb{R}$ will be denoted by $\mathcal{R}(\Omega)$.

The Riemann integral may be extended to unbounded functions and unbounded intervals by taking the limit of the integral using a suitably truncated sequence of functions f_n. The resulting integrals are called *improper integrals*. The reader should be familiar with the many properties of the Riemann integral, such as linearity, monotonicity, and the anti-derivative property which makes it indispensable in many areas of mathematics and its applications. Its main drawback is that $\mathcal{R}$ is not closed under some limiting operations, as the following examples show.

For every $n \in \mathbb{N}$, let $f_n : [0, 1] \to \mathbb{R}$ be defined by

$$f_n(x) = \begin{cases} 1, & x = p/q \in \mathbb{Q} \cap [0, 1], \ q < n \\ 0, & \text{otherwise}, \end{cases} \tag{A.2}$$

where p and q have no common factors. f_n is clearly integrable for every n, since it is continuous except at a finite number of points, and its integral is 0 for every n, so $\lim \int_0^1 f_n(x)\,dx = 0$. On the other hand, the sequence (f_n) increases with n and converges as $n \to \infty$ to the non-integrable Dirichlet function f defined in Eq. (A.1).

Furthermore, even when $\lim f_n$ is integrable, the two limits, $\lim \int f_n(x)\, dx$ and $\int \lim f_n(x)\, dx$, may not be equal, as the sequence of functions $nx\left(1 - x^2\right)^n$ on $[0, 1]$ demonstrates (see Exercise A.1)

These deficiencies in the Riemann integral do not place important limitations in practical problems that we have to deal with, but they do matter when we try to build a coherent mathematical theory for the inner product space in Chap. 1, for example (see Sect. 1.4). To minimize these deficiencies several attempts were made to develop an alternative theory of integration in the late nineteenth and early twentieth century. The theory proposed by Henri Lebesgue (1875–1941) received wide acceptance as it generalizes Riemann integration and expands the class of integrable functions so as to mitigate these deficiencies, and we now give a brief outline of this theory.

A.1   Lebesgue Measure

In Riemann integration a central role is played by the *measure* of a finite interval I from a to b, defined by its length $l(I) = b - a \geq 0$ regardless of whether the interval is closed, half-closed, open, or even a single point (where $a = b$). If I is unbounded then $l(I) = \infty$. With the Dirichlet function we have no method for measuring the "length" of such subsets of $\mathbb{R}$ as $[0, 1] \cap \mathbb{Q}$ or $[0, 1] \cap \mathbb{Q}^c$, where $\mathbb{Q}^c$, the complement of $\mathbb{Q}$, denotes the irrational real numbers. So our first task is to extend the concept of length from intervals to more general subsets of $\mathbb{R}$. In essence we would like to define a non-negative function m on the subsets of $\mathbb{R}$ such that

1. $m(I) = l(I)$ for any interval in $\mathbb{R}$.

2. If E_1 and E_2 are disjoint subsets of $\mathbb{R}$, then $m(E_1 \cup E_2) = m(E_1) + m(E_2)$. More generally, if (E_i) is a sequence of pairwise disjoint subsets of $\mathbb{R}$, then

$$m\left(\bigcup_{i=1}^{\infty} E_n\right) = \sum_{i=1}^{\infty} m(E_n).$$

This property is called *countable additivity.*

3. m is invariant under translation, that is, $m(E + x) = m(E)$ for any $x \in \mathbb{R}$, $E \subseteq \mathbb{R}$.

It turns out that there is no function which satisfies these three conditions on all subsets of $\mathbb{R}$ [14]. But since they are basic for any definition of measure, we shall proceed to construct a measure function on the largest class of subsets of $\mathbb{R}$ which satify these three conditions and includes all the intervals. For any subset E of $\mathbb{R}$, an *open cover* of E is any collection (finite or infinite) of open sets $G_\lambda \subseteq \mathbb{R}$ such that $E \subseteq \bigcup_{\lambda \in \Lambda} G_\lambda$. We should recall that a set $G \subseteq \mathbb{R}$ is *open* (in $\mathbb{R}$) if ever point $x \in G$ has a neighbourhood $(x - \varepsilon, x + \varepsilon)$, for some $\varepsilon > 0$, which lies in G. G is *closed* if its complement $G^c = \mathbb{R} \backslash G$ is open.

Definition A.1 For any $E \subseteq \mathbb{R}$, we define the *outer measure* of E as

$$m^* (E) = \inf_{E \subseteq \cup I_i} \sum_{i=1}^{\infty} l (I_i), \tag{A.3}$$

where the infimum is taken over all coverings of E by collections $\{I_i : i \in \mathbb{N}\}$ of open real intervals.

Thus m^* is a non-negative function defined on all subsets of $\mathbb{R}$. Let $J = [a, b]$ be any closed and bounded interval. The interval $I = (a - \varepsilon, b + \varepsilon)$ covers J for any $\varepsilon > 0$, therefore

$$m^* (J) \leq l (I) = b - a + 2\varepsilon.$$

$\varepsilon > 0$ being arbitrary, we conclude that $m^* (J) \leq b - a$. Suppose $\cup I_i$ is any other open cover of J. Since J is compact, i.e. closed and bounded, then, by the Heine-Borel theore, it has a finite subcover, say $\bigcup_{i=1}^{n} J_i$, of open intervals. Renumbering $\{J_i\}$ if necessary, we may assume, without loss of generality, that the intervals $J_i = (a_i, b_i)$ are arranged from left to right with the endpoints of J_1 and J_n satisfying $a_1 < a < b_1 < \cdots < a_n < b < b_n$, as shown in Fig. A.1 with $a_i < b_{i-1}$ for all $2 \leq i \leq n$. Consequently

$$\sum_{i=1}^{\infty} l (I_i) \geq \sum_{i=1}^{n} l (J_i) = \sum_{i=1}^{n} (b_i - a_i) > b_n - a_1 > b - a.$$

Since $b - a$ is a lower bound for $\sum l (I_i)$ for all open coverings $\bigcup I_i$ of J, this implies $m^* (J) \geq b - a$. Hence $m^* (J) = b - a$.

Similarly we easily see that $m^* (\varnothing) = 0$ and $m^* (\{x\}) = 0$ for any real number x. We can also prove that outer measure is monotonic and countably subadditive (see [1], that is,

$$E \subseteq F \Rightarrow m^* (E) \leq m^* (F) \quad \text{for all } E, F \subseteq \mathbb{R},$$

$$m^* \left(\bigcup_{i=1}^{\infty} E_i \right) \leq \sum_{i=1}^{\infty} m^* (E_i) \quad \text{for all } E_i \subseteq \mathbb{R}.$$

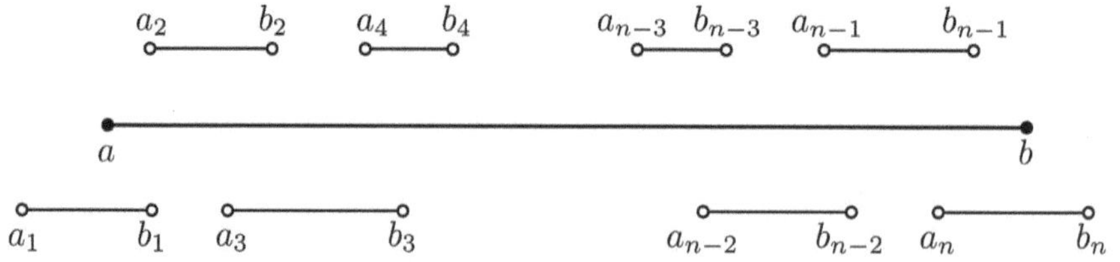

Fig. A.1 Cover of $[a, b]$ by open intervals

From this we conclude that the outer measure of any real interval from a to b is its length $b - a$, and that the outer measure of any countable set of real numbers is zero. In particular $m^*(\mathbb{Q}) = 0$ and $m^*(\mathbb{R}) = \infty$. We can also show that $m^*(E + c) = m^*(E)$ and $m^*(cE) = |c|\, m^*(E)$ for any $E \subseteq \mathbb{R}$ and $c \in \mathbb{R}$ [1]. But we know from an earlier observation that m^* is not countably additive unless we restrict m^* to certain subsets of $\mathbb{R}$. This we do in the following definition.

Definition A.2 A set $E \subseteq \mathbb{R}$ is said to be *measurable* if

$$m^*(A) = m^*(A \cap E) + m^*\left(A \cap E^c\right) \quad \text{for all } A \subseteq \mathbb{R}.$$

The measurable sets in $\mathbb{R}$ will be denoted by $\mathcal{M}$.

Since we always have $m^*(A) \leq m^*(A \cap E) + m^*(A \cap E^c)$ we need only show that $m^*(A) \geq m^*(A \cap E) + m^*(A \cap E^c)$ for any A in order to prove that E is measurable. From the definition it is clear that E^c is measurable whenever E is, and that both $\varnothing$ and $\mathbb{R}$ are measurable. Furthermore, if $m^*(E) = 0$, then by monotonicity $m^*(A \cap E) = 0$ and we obtain $m^*(A) \geq m^*(A \cap E^c) = m^*(A \cap E^c) + m^*(A \cap E)$. So any subset of $\mathbb{R}$ whose outer measure is 0 is measurable.

The property that remains to be verified is that $\mathcal{M}$ is countably additive. We can show that, if E_1 and E_2 are measurable, then so is $E_1 \cup E_2$ by directly applying Definition A.2. Using induction we can extend this result to any finite union of measurable sets. Therefore $\mathcal{M}$ is closed under complementation and the formation of finite unions. To extend this to countable unions and verify that

$$m^*\left(\bigcup_{i=1}^{\infty} E_i\right) = \sum_{i=1}^{\infty} m^*(E_i)$$

for any disjoint collection $\{E_i : i \in \mathbb{N}\}$ of measurable sets, we refer the reader to [1] or [14].

The restriction of m^* to $\mathcal{M}$ will henceforth be the measure function

$$m = m^*\big|_{\mathcal{M}} : \mathcal{M} \to [0, \infty],$$

known as *Lebesgue measure,* which is the natural extension of l from the intervals in $\mathbb{R}$ to the Lebesgue measurable sets $\mathcal{M}$. We have just seen that $\mathcal{M}$ is closed under countable unions and complementation. Consequently it is also closed under countable intersections. Such a class of subsets of $\mathbb{R}$ is called a σ-*algebra* on $\mathbb{R}$.

It can be shown that for any class $\mathcal{C}$ of subsets of $\mathbb{R}$ there is a smallest σ-algebra on $\mathbb{R}$ that contains $\mathcal{C}$ (see [1, 14]), which will be denoted by $\mathcal{A}(\mathcal{C})$, and refered to as the σ-algebra *generated* by $\mathcal{C}$. The σ-algebra generated by the finite open intervals in $\mathbb{R}$ is called the *Borel* σ-algebra on $\mathbb{R}$, and denoted by $\mathcal{B}$. It includes all the interesting subsets of $\mathbb{R}$ that we have occasion to deal with, such as intervals, open sets, closed sets, and countable sets. In fact the Borel σ-algebra can also be generated by the closed intervals as well as the semi-infinite intervals. It can be shown that Borel sets are measurable, and hence $\mathcal{B} \subset \mathcal{M}$.

To define measurability for functions we recall how continuity for a function $f : \Omega \to \mathbb{R}$ was defined by the topological condition that the inverse image $f^{-1}(U)$ is open in Ω whenever U is open in $\mathbb{R}$. Let $\Omega \subseteq \mathbb{R}$ be a measurable set. For the function f to be measurable, a corresponding definition would state that $f^{-1}(U)$ be measurable in Ω whenever U is open in $\mathbb{R}$, and that is one definition of measurability [15]. But since the open sets in $\mathbb{R}$ generate the *Borel* σ-algebra $\mathcal{B}$ by countable unions and intersections, an equivalent definition would be that f is measurable if $f^{-1}(B)$ is measurable for every Borel set B in $\mathbb{R}$. Note that the fact that $\mathcal{B} \subset \mathcal{M}$ insures that every continuous function is measurable. Another equivalent definition, which is commonly used and simpler to apply, relies on the fact that semi-infinite intervals (α, ∞) also generate $\mathcal{B}$ (see [14]), so we can state that

$$f \text{ is measurable if, for all } \alpha \in \mathbb{R}, \{x \in \Omega : f(x) > \alpha\} \text{ is measurable.} \qquad \text{(A.4)}$$

For an extended real function $f : \Omega \to \overline{\mathbb{R}} = \mathbb{R} \cup \{\infty, -\infty\}$ to be measurable the sets $f^{-1}(\infty)$ and $f^{-1}(-\infty)$ must also be measurable.

Applying the criterion (A.4) above to the Dirichlet function in (A.1), we see that

$$f^{-1}(\alpha, \infty) = \begin{cases} [0, 1] & \text{if } \alpha < 0 \\ \mathbb{Q} & \text{if } 0 \le \alpha < 1 \\ \varnothing & \text{if } \alpha > 1. \end{cases}$$

Because $[0, 1]$, $\mathbb{Q}$, and $\varnothing$ are all measurable sets, we conclude that this function is measurable. Note that $f(x) = 0$ except when $x \in \mathbb{Q}$, and that $m(\mathbb{Q}) = 0$. Such a function, which is 0 on its domain except for a subset of measure 0, is said to be 0 *almost everywhere* (abbreviated to "a.e"). Consequently we say that two functions are "equal a.e." if their difference is 0 a.e. Any linear combination or product of two measurable functions is also measurable [1].

A.2   Lebesgue Integration

For any set $E \subseteq \mathbb{R}$, the *characteristic function* of E is defined as

$$\chi_E(x) = \begin{cases} 1, & x \in E \\ 0, & x \notin E, \end{cases}$$

and χ_E is measurable if, and only if, E is measurable. Given any finite collection $\{E_1, E_2, \cdots, E_n\}$ of disjoint, measurable sets in $\mathbb{R}$, the linear combination

$$\varphi = \sum_{i=1}^{n} c_i \chi_E, \quad c_i \in \mathbb{R} \qquad \text{(A.5)}$$

is called a *simple function,* whose representation is unique if $c_i \neq c_j$ whenever $i \neq j$. It is measurable with domain $\bigcup_{i=1}^{n} E_i$ and its range is the finite set $\{c_i : i = 1, 2, \cdots, n\}$.

The significance of the simple functions is that any non-negative measurable function can be approached from below by a sequence of increasing simple functions. That is to say, if $f : \Omega \rightarrow [0, \infty]$ is measurable, then there is a sequence (φ_n) of non-negative simple functions such that $\varphi_{n+1} \geq \varphi_n$ for all n and $\lim_{n \to \infty} \varphi_n(x) = f(x)$ for every $x \in \Omega$, expressed symbolically by $\varphi_n \nearrow f$ [1].

The Lebesgue integral of the non-negative simple function in Eq. (A.5) is defined as

$$\int_{\cup E_i} \varphi dm = \sum_{i=1}^{n} c_i m(E_i),$$

which agrees with our notion of the integral of a function as the "area under the curve" over the domain of the function. In case $m_i(E_i) = \infty$ and $c_i = 0$ for some i, we set $c_i m(E_i) = 0$. Let $S(\Omega)$ denote the set of simple functions defined on the measurable set $\Omega \subseteq \mathbb{R}$, and let $S_+(\Omega)$ be its non-negative subset. We define the integral of a non-negative measurable function f on Ω, $f : \Omega \rightarrow [0, \infty]$, as

$$\int_{\Omega} f dm = \sup \left\{ \int_{\Omega} \varphi(x) dm : \varphi \in S_+(\Omega), \varphi \leq f \right\}.$$

Since there is a sequence (φ_n) in S_+ such that $\varphi_n \nearrow f$, we also have

$$\lim_{n \to \infty} \int_{\Omega} \varphi_n dm = \int_{\Omega} f dm,$$

which provides a more practical method for computing $\int_{\Omega} f dm$. The non-negative measureable function f is said to be *integrable* (in the sense of Lebesgue) if the integral $\int_{\Omega} f dm$ is finite.

This same definition of the integral extends to a general measurable function $f : \Omega \rightarrow \overline{\mathbb{R}}$ since such a function can be expressed as the difference of two non-negative measrrurable functions, $f = f^+ - f^-$, where f^+ is the positive part of f and f^- is its negative part. In this case we say f is integrable if both f^+ and f^- are integrable, and the *Lebesgue integral* of f is then defined as $\int_{\Omega} f dm = \int_{\Omega} f^+ dm - \int_{\Omega} f^- dm$. It follows that $|f| = f^+ + f^-$ is also integrable. Lebesgue integrable functions on Ω will be denoted by $\mathcal{L}^1(\Omega)$, which is a proper subset of the measurable functions on Ω. Consequently, for the function $f = f^+ - f^-$ to be in $\mathcal{L}^1(\Omega)$ both f^+ and f^- have to be in $\mathcal{L}^1(\Omega)$, and this of course means $|f| = f^+ + f^-$ also lies in $\mathcal{L}^1(\Omega)$.

Properties such as linearity and monotonicity also hold for Lebesgue integrals just as they do for Riemann integrals. The main advantage of Lebesgue integration, from our point of view, is that it expands the class $\mathcal{R}$ of Riemann integrable functions to the point of providing some closure under limiting operations in the resulting space $\mathcal{L}^1$ which is missing in $\mathcal{R}$. It also has special properties peculiar to the role that measure plays in its definition, and whose proofs may be looked up in several standard references. Here are some of the more significant properties from our perspective:

1. If $f \in \mathcal{L}^1(\Omega)$ and $g = f$ a.e. on Ω, then $g \in \mathcal{L}^1(\Omega)$ and

$$\int_\Omega g\,dm = \int_\Omega f\,dm.$$

   In other words, each "function" in $\mathcal{L}^1(\Omega)$ actually represents an infinite class of functions which are equal almost everywhere.
2. If $f \in \mathcal{L}^1(\Omega)$ then $m\{x \in \Omega : |f(x)| = \infty\} = 0$. This allows us to assume that any integrable function does not assume the values $\pm\infty$. Hence we can always write $\int_{\overline{\mathbb{R}}} f\,dm = \int_{\mathbb{R}} f\,dm$ when f is integrable. By the same token the Lebesgue integral over any open interval equals the corresponding integral over its closure [14].
3. If f is a bounded Riemann integrable function on a bounded interval I from a to b, i.e. excluding improper integrals, then $f \in \mathcal{L}^1(I)$ and the two integrals are equal,

$$\int_I f\,dm = \int_a^b f(x)\,dx.$$

   This is an important result which allows us to use all the techniques of Riemann integration at our disposal to evaluate Lebesgue integrals of bounded functions over bounded intervals. In particular, if $f \in \mathcal{L}^1(a, b)$ and $F(x) = \int_{[a,x]} f\,dm = \int_a^x f(x)\,dx$ is the primitive function of f, then

$$F'(x) = f(x) \quad \text{for all } x \in [a, b]$$

   provided f is continuous on $[a, b]$, otherwise the equality is a.e. (see [14]). This result is equally useful in that it allows us to use the theorems of Lebesgue integration, some of which are mentioned below, where they apply when dealing with Riemann integrals.
4. A bounded function f on a bounded interval is Riemann integrable if, and only if, it is continuous almost everywhere. This is another important result in $\mathcal{R}$ which could not have been proven (or even visualized) without the concept of measure. See [1] for the proof. The Dirichlet function mentioned earlier, for example, is not continuous on the set of irrational numbers in $[0, 1]$ (see Exercise A.4) which has measure 1, so it cannot be Riemann integrable.

5. If the improper integral $\int_a^\infty |f(x)|\,dx$ exists then both $\int_a^\infty f(x)\,dx$ and the Lebesgue integral $\int_{[a,\infty)} f\,dm$ exist and are equal. But the existence of $\int_a^\infty f(x)\,dx$ does not guarantee that $f \in \mathcal{L}^1(0,\infty)$, i.e. that $\int_{[a,\infty)} |f|\,dm$ exists. As an example, consider $f(x) = (\sin x)/x$ (Exercise 1.28).

6. Let (f_n) be a sequence of measurable functions on Ω and $f_n \to f$ a.e. If there is a function $g \in \mathcal{L}^1(\Omega)$ such that $|f_n(x)| \leq g(x)$ a.e. for every $n \in \mathbb{N}$, then $f \in \mathcal{L}^1(\Omega)$ and

$$\lim_{n\to\infty} \int_\Omega f_n dm = \int_\Omega f\,dm.$$

This is known as the *dominated convergence theorem*, which is one of the fundamental theorems of Lebesgue integration.

7. Let (f_n) be a sequence of measurable functions on Ω, and suppose that there is a real number M such that $|f_n(x)| \leq M$ for all n and x. If $m(\Omega) < \infty$ and $f_n \to f$ a.e, then $f \in \mathcal{L}^1(\Omega)$ and

$$\lim_{n\to\infty} \int_\Omega f_n dm = \int_\Omega f\,dm.$$

This is another important result, known as the *bounded convergence theorem*. It follows directly from the dominated convergence theorem by taking the upper bound of $|f_n(x)|$ on Ω, which is a constant, as the dominating function. Neither of these two theorems can be proved within the theory of Riemann integration. With $\Omega = [0, 1]$ and $M = 1$, the Lebesgue integrability of the Dirichlet function defined in (A.1) follows directly from the bounded convergence theorem.

8. If $f \in \mathcal{L}^1(\mathbb{R})$, then

$$\lim_{n\to\infty} \int_{\mathbb{R}} f(x)\cos nx\,dx = 0. \tag{A.6}$$

This is known as the *Riemann-Lebesgue Lemma*, and it plays a significant role in Fourier analysis. Its proof relies on selecting a sequence of simple functions (φ_n) which approach f such that each φ_n is supported in a compact interval I_n, i.e. $\varphi_n = 0$ outside I_n (see [1]).

In this section we have used two slightly different notations in order to distinguish between the Riemann and the Lebesgue integrals. But, in view of the fact that proper Riemann integrable functions are a subset of $\mathcal{L}^1$ (as stated in property 3 above), we do not always have to make this distinction, and we choose to use the notation $\int_a^b f(x)\,dx$ or $\int_E f(x)\,dx$ to designate either one.

Exercises

A.1 Given that $g_n(x) = nx \left(1 - x^2\right)^n$ for all $x \in [0, 1]$, compute the limits for the Riemann integrals

$$\int_0^1 \lim_{n \to \infty} g_n(x)\, dx \quad \text{and} \quad \lim_{n \to \infty} \int_0^1 g_n(x)\, dx$$

to show that they are not equal.

A.2 If the function $f : \Omega \to \overline{\mathbb{R}}$ is measurable, show that cf for any $c \in \mathbb{R}$ is also measurable. Is the converse of this statement true, i.e. if cf is measurable can we conclude that f is also measurable?

A.3 Given that the function $f : \Omega \to \overline{\mathbb{R}}$ is measurable, prove that $|f|$ is measurable. What can we say about the converse of this statement?

A.4 Show in detail how the Dirichlet function defined by Eq. (A.1) fails to meet the condition of Riemann integrability as stated in property 5 above, namely continuity almost everywhere on $[0, 1]$.

A.5 Use the dominated convergence theorem to evaluate

$$\lim_{n \to \infty} \int_0^n \left(1 - \frac{x}{n}\right)^n e^{-x} dx.$$

Appendix B
Solutions to Selected Exercises

Chapter 1

1.2 (a) complex vector space, (b) real vector space, (c) not a vector space, (d) real vector space.

1.4 Assume that $\{x_1, \cdots, x_n\}$ and $\{y_1, \cdots, y_m\}$ are bases of the same vector space. If $m > n$, express each y_i, $1 \le i \le n$, as a linear combination of $x_1, \cdots, x_n$ and show that this leads to a contradiction. The resulting system of n linear equations can be solved uniquely for each x_i, $1 \le i \le n$, as a linear combination of y_i, $1 \le i \le n$. Since each vector $y_{n+1}, \cdots, y_m$ is also a linear combination of $x_1, \cdots, x_n$ (and hence of $y_1, \cdots, y_n$) this contradicts the linear independence of $\{y_1, \cdots, y_m\}$. Similarly if $m < n$.

1.7 Recall that a determinant is zero if, and only if, one of its rows (or columns) is a linear combination of the other rows (or columns).

1.8 Use the equality $\|\mathbf{x} + \mathbf{y}\|^2 = \|\mathbf{x}\|^2 + 2\mathrm{Re}\,\langle \mathbf{x}, \mathbf{y}\rangle + \|\mathbf{y}\|^2$. Consider $\mathbf{x} = (1, 1)$ and $\mathbf{y} = (i, i)$.

1.10 (a) 0, (b) 2/3, (c) 8/3, (d) $\sqrt{14}$.

1.12 $\langle f, f_1\rangle / \|f_1\| = \sqrt{\pi/2}$, $\langle f, f_2\rangle / \|f_2\| = 0$, $\langle f, f_3\rangle / \|f_3\| = \sqrt{\pi}/2$.

1.14 Evaluate a linear combination at $x = 0, \pm 1$. No, because $|x| = -x$ on $[-1, 0]$..

1.17 $a = -1$, $b = 1/6$.

1.21 Pointwise. Both sides equal 1.

1.22 Pointwise convergent to

$$f(x) = \begin{cases} 0, & x = 0 \\ 1 - x, & 0 < x \le 1. \end{cases}$$

1.24 $0 \le \dfrac{x}{n + x} \le \dfrac{a}{n} \overset{u}{\to} 0$. On $[0, \infty)$ consider the sequence $f_n(x) = n$.

1.25 The interval is unbounded.

© The Author(s), under exclusive license to Springer-Verlag London Ltd., part of Springer Nature 2026
M. A. Al-Gwaiz, *Sturm-Liouville Theory and its Applications*, Springer Undergraduate Mathematics Series, https://doi.org/10.1007/978-1-4471-7610-7

1.27 (a) Uniform convergence on $\mathbb{R}$. (b) Converges on $(-1, 1)$, diverges outside outside $(-1, 1)$.

1.31 Use Corollary 1.16 to prove uniform convergence of the series and its derivative in $[-R + \varepsilon, R - \varepsilon]$ for any $\varepsilon > 0$.

1.36 Use the definition of the Riemann integral to show that f is not integrable.

1.37 (i) $1/\sqrt{2}$, (ii) not in $\mathcal{L}^2 (0, \infty)$, (iii) 1, (iv) not in $\mathcal{L}^2 (0, \infty)$.

1.38 If either f or g is 0 we have linear dependence, otherwise show that
$$\| f/ \|f\| - g/ \|g\| \|^2 = 0.$$

1.42 If $f(x) \not\to 0$ show that $\|f\| = \infty$.

1.45 $\sin^3 x = \dfrac{3}{4} \sin x - \dfrac{1}{4} \sin 3x$.

1.47 Use the monotonic property of the integral.

1.48 (a) 1, (b) 1, (c) 0.

1.49 (a) convergent, (b) convergent, (c) divergent.

1.51 Use the triangle inequality.

1.57 $c_1 = 1$, $c_2 = -2/\pi$, $c_3 = -1/\pi$.

1.58 $a_0 = \pi/2$, $a_1 = -4/\pi$, $a_2 = 0$, $b_1 = 0$, $b_2 = 0$.

1.61 $a_n = 1/n$. $\mathcal{L}^2$ convergence.

Chapter 2

2.1

(a) $y = e^{2x}(c_1 \cos \sqrt{3}x + c_2 \sin \sqrt{3}x) + e^x/4$.

(b) $y = c_1 x^2 + c_2 + x^3$.

(c) $y = x^{-1}(c_1 + c_2 \log x) + \dfrac{1}{4}x - 1$.

2.3 $c_{n+2} = -\dfrac{2}{n+1} c_n$, $y = c_0 \left(1 - 2x^2 + \frac{4}{3}x^4 + \cdots \right) + c_1 \left(x - x^3 + \frac{1}{2}x^5 + \cdots \right)$, $x \in \mathbb{R}$. The series converges for all $x \in \mathbb{R}$ since q and r are analytic in $\mathbb{R}$.

2.5 A second order equation has at most two linearly independent solutions.

2.7 (a) $y'' + 2y' + 5y = 0$, (b) $x^2 y'' + xy' - y = 0$, (c) $xy'' + y' = 0$.

2.8 Use Lemma 2.7 and the fact that a bounded infinite set of real numbers has at least one cluster point. This property of the real (as well as the complex) numbers is known as the *Bolzano-Weierstrass theorem*. Being isolated, the zeros cannot have a finite cluster point.

2.12 The solutions of (a) and (c) are oscillatory.

2.13 $y = x^{-1/2}(c_1 \cos x + c_2 \sin x)$. The zeros of $x^{-1/2} \cos x$ are $\{\frac{\pi}{2} + n\pi\}$, and those of $x^{-1/2} \sin x$ are $\{n\pi\}$, $n \in \mathbb{N}_0$.

2.15 Consider $y'' + xy = 0$ on the two interval $(-\infty, -\varepsilon]$ and $[\varepsilon, \infty)$ for any $\varepsilon > 0$, and use the Sturm comparison theorem.

2.17 (a) $e^{\pm\sqrt{\lambda}x}$ when $\lambda \in \mathbb{C}\backslash \{0\}$, and $\{1, x\}$ when $\lambda = 0$ (b) $e^{-\sqrt{\lambda}x}$, $\mathrm{Re}\sqrt{\lambda} > 0$.

2.19 (a) $\rho = 1/x^2$, (c) $\rho = e^{-x^3/3}$.

2.21 $\rho = e^{2x}$, $\lambda_n = n^2\pi^2 + 1$, $u_n(x) = e^{-x} \sin n\pi x$

2.23 $\lambda_n = n^2\pi^2(b-a)^{-2}$, $u_n(x) = \sin\left(\dfrac{n\pi(x-a)}{b-a}\right)$, $n \in \mathbb{N}$.

2.27 (a), (b), (c) and (f).

2.29 Change the independent variable to $\xi = x + 3$ and solve. $\lambda_n = \left(\dfrac{n\pi}{\log 4}\right)^2 + \dfrac{1}{4}$, $y_n(x) = (x+3)^{-1/2} \sin\left(\dfrac{n\pi}{\log 4}\log(x+3)\right)$, $n \in \mathbb{N}$.

2.31 Refer to Example 2.16. $\lambda_n = \left(\dfrac{\alpha_n}{l}\right)^2$, $u_n(x) = \dfrac{\alpha_n}{l}\cos\left(\dfrac{\alpha_n}{l}x\right) + \sin\left(\dfrac{\alpha_n}{l}x\right)$, $n \in \mathbb{N}$.

2.32 Multiply by $\bar{u}$ and integrate over $[a, b]$.

Chapter 3

3.2 No, because its sum is discontinuous at $x = 0$ (Example 3.4).

3.4 $\pi - |x| = \dfrac{\pi}{2} + \dfrac{4}{\pi}\sum_{n=0}^{\infty}\dfrac{1}{(2n+1)^2}\cos(2n+1)x$. Uniformly convergent by the M-test.

3.6 Use the M-test.

3.9 (a) and (c) are piecewise continuous. (b) and (d) are piecewise smooth.

3.11 Use the definition of the derivative at $x = 0$ to show that $f'(0)$ exists, then show that $\lim_{x\to 0+} f'(x)$ does not exist.

3.15

(a) $f(x) = 2\sum_{n=1}^{\infty}\dfrac{(-1)^{n+1}}{n}\sin nx$.

(b) $f(x) = \dfrac{1}{2} - \dfrac{4}{\pi^2}\sum_{k=0}^{\infty}\dfrac{1}{(2k+1)^2}\cos(2k+1)\pi x$.

(c) $f(x) = \dfrac{1}{2} - \dfrac{1}{2}\cos 2x$.

(d) $f(x) = 2l^3\sum_{n=1}^{\infty}(-1)^n\left(\dfrac{6}{n^3\pi^3} - \dfrac{1}{n\pi}\right)\sin\left(\dfrac{n\pi x}{l}\right)$.

3.16 Convergence of the infinite series is uniform in (b) only.

3.17 $x = \pm l$ are points of discontinuity for f and continuity for g, hence the sum of the Fourier series at these points is $f(l^+) - f(l^-) = 0$ for the odd function f and $g(l) = l^3$ for the even function g.

3.19 Setting $x = 0$ in the expansion yields $\pi^2 = 8\sum_{n=0}^{\infty}(2n+1)^{-2}$.

3.21 $x^2 = \dfrac{\pi^2}{3} + 4\sum_{n=1}^{\infty}\dfrac{(-1)^n}{n^2}\cos nx$. Evaluate at $x = 0$ and $x = \pi$ for (a) and (b).

3.23 f' is an odd function which is periodic in π defined as $f'(x) = \cos x$ on $[0, \pi]$, hence $S(x) = \sum_{k=1}^{\infty} b_k \sin kx$ where $b_k = \dfrac{2k}{\pi}\left[1 + (-1)^k\right]\big/\left(k^2 - 1\right)$, $k > 1$, $b_1 = 0$, and $S(\pi/2 + n\pi) = S(n\pi) = 0$ for all n.

3.27 $u(x, t) = \dfrac{3}{4} e^{-t} \sin x - \dfrac{1}{4} e^{-9t} \sin 3x.$

3.29 $u(x, t) = \displaystyle\sum_{n=1}^{\infty} a_n \sin \dfrac{n\pi}{l} x \cos \dfrac{n\pi}{l} t,\ a_n = \dfrac{2}{l} \displaystyle\int_0^l x(l - x) \sin \dfrac{n\pi}{l} x\, dx.$

3.31 Assume $u(x, t) = v(x, t) + \psi(x)$ where v satisfies the homogeneous wave equation with homogeneous boundary conditions at $x = 0$ and $x = l$. This leads to $\psi(x) = \dfrac{g}{2c^2}(x^2 - lx)$, and $v(x, t) = \displaystyle\sum_{n=1}^{\infty} a_n \sin \dfrac{n\pi}{l} x \cos \dfrac{cn\pi}{l} t$, with

$$a_n = -\frac{2}{l} \int_0^l \psi(x) \sin \frac{n\pi}{l} x\, dx.$$

3.33 Assume $u(x, y, t) = v(x, y)w(t)$ and conclude that $w''/w = c^2 \Delta v/v = -\lambda^2$. Hence $w(t) = A \cos \lambda t + B \sin \lambda t$. Then assume $v(x, y) = X(x)Y(y)$, and use the given boundary conditions to conclude that

$$\lambda_{mn} = \sqrt{\frac{n^2}{a^2} + \frac{m^2}{b^2}}\, \pi, \quad m, n \in \mathbb{N},$$

$$X(x) = \sin \frac{n\pi}{a} x, \quad Y(y) = \sin \frac{m\pi}{b} y,$$

$$u_{mn}(x, y, t) = (A_{mn} \cos \lambda_{mn} ct + B_{mn} \sin \lambda_{mn} ct) \sin \frac{n\pi}{a} x \, \sin \frac{m\pi}{b} y.$$

Apply the initial conditions to the solution to evaluate the coefficients A_{mn}, B_{mn}.

3.35 $u(x, y) = \left(\sinh \dfrac{3\pi}{2} \right)^{-1} \sin \dfrac{3\pi}{2} x \, \sinh \dfrac{3\pi}{2} y.$

3.37

(a) $u_n(r, \theta) = \begin{cases} a_0 + d_0 \log r, & n = 0. \\ (c_n r^n + d_n r^{-n})(a_n \cos n\theta + b_n \sin n\theta), & n \in \mathbb{N}. \end{cases}$

(b) Inside the circle: $u(r, \theta) = A_0 + \displaystyle\sum_{n=1}^{\infty} \left(\dfrac{r}{R} \right)^n (A_n \cos n\theta + B_n \sin n\theta).$

(c) Outside the circle: $u(r, \theta) = A_0 + \displaystyle\sum_{n=1}^{\infty} \left(\dfrac{R}{r} \right)^n (A_n \cos n\theta + B_n \sin n\theta).$

Chapter 4

4.3 From the recursion formula (4.7) with $k = 2j$, it follows that

$$\lim_{j \to \infty} \frac{\left| c_{2(j+1)} x^{2(j+1)} \right|}{\left| c_{2j} x^{2j} \right|} = x^2 < 1 \quad \text{for all } x \in (-1, 1).$$

The same conclusion holds if $k = 2j + 1$.

4.5 The first two formulas follow from the fact that P_n is an even function when n is even, and odd when n is odd.

$$P_{2n}(0) = a_0 = \frac{(-1)^n (2n)!}{2^{2n} n! n!} = (-1)^n \frac{(2n-1)\cdots(3)(1)}{(2n)\cdots(4)(2)}.$$

4.7 Using Rodrigues' formula, first show that

$$P_{n+1}(x) = \frac{1}{2^{n+1}(n+1)!} \frac{d^{n+1}}{dx^{n+1}} \left(x^2 - 1\right)^{n+1}$$

$$= \frac{2n+1}{2^n n!} \frac{d^{n-1}}{dx^{n-1}} \left(x^2 - 1\right)^n + P_{n-1}(x),$$

then differentiate both sides to obtain $P_n(x) = \left[P'_{n+1}(x) - P'_{n-1}(x)\right]/(2n+1)$ and the first formula. The second follows from the relation $P_n(\pm 1) = (\pm 1)^n$.

4.11

(a) $1 - x^3 = P_0(x) - \frac{3}{5} P_1(x) - \frac{2}{5} P_3(x).$

(b) $|x| = \frac{1}{2} P_0(x) + \frac{5}{8} P_2(x) - \frac{3}{16} P_4(x) + \cdots .$

4.13 $f(x) = \sum_{n=0}^{\infty} c_n P_n(x)$, where $c_n = \frac{2n+1}{2} \int_{-1}^{1} f(x) P_n(x) dx$. Since f is odd, $c_n = 0$ for all even values of n. For $n = 2k + 1$,

$$c_{2k+1} = (4k+3) \int_0^1 P_{2k+1}(x) dx$$

$$= (4k+3) \frac{1}{4k+3} [P_{2k}(0) - P_{2k+2}(0)]$$

$$= (-1)^k \left[\frac{(2k)!}{2^{2k} k! k!} + \frac{(2k+2)!}{2^{2k+2}(k+1)!(k+1)!} \right]$$

$$= (-1)^k \frac{(2k)!}{2^{2k} k! k!} \frac{(4k+3)}{(2k+2)}, \quad k \in \mathbb{N}_0.$$

Hence $f(x) = \frac{3}{2} P_1(x) - \frac{7}{8} P_3(x) + \frac{11}{16} P_5(x) + \cdots$. At $x = 0$ the sum is 0.

4.15 $\left(\int_{-\infty}^{\infty} e^{-x^2} dx\right)^2 = 4 \int_0^{\infty} \int_0^{\infty} e^{-(x^2+y^2)} dx dy = 4 \int_0^{\pi/2} \int_0^{\infty} e^{-r^2} r \, dr d\theta = \pi.$

4.16 Replace t by $-t$ in Eq. (4.25) to obtain

$$\sum_{n=0}^{\infty} \frac{1}{n!} H_n(x)(-t)^n = e^{-2xt - t^2} = \sum_{n=0}^{\infty} \frac{1}{n!} H_n(-x) t^n,$$

which implies $H_n(-x) = (-1)^n H_n(x)$.

4.17 Setting $x = 0$ in (4.25) yields

$$\sum_{k=0}^{\infty} \frac{1}{k!} H_k(0) t^k = e^{-t^2} = \sum_{n=0}^{\infty} (-1)^n \frac{1}{n!} t^{2n}.$$

By equating corresponding coefficients we obtain the desired formulas.

4.19 If $m = 2n$,

$$x^{2n} = \frac{(2n)!}{2^{2n}} \sum_{k=0}^{n} \frac{H_{2k}(x)}{(2k)!(n-k)!}.$$

If $m = 2n + 1$,

$$x^{2n+1} = \frac{(2n+1)!}{2^{2n+1}} \sum_{k=0}^{n} \frac{H_{2k+1}(x)}{(2k+1)!(n-k)!}, \quad x \in \mathbb{R}, \ n \in \mathbb{N}_0.$$

4.21 Use Leibnitz' rule for the derivative of a product,

$$(fg)^{(n)} = \sum_{k=0}^{n} \binom{n}{k} f^{(n-k)} g^{(k)},$$

with $f(x) = x^n$ and $g(x) = e^{-x}$.

4.22 $x^3 - x = 5L_0(x) - 17L_1(x) + 18L_2(x) - 6L_3$.

4.23 $x^m = \sum_{n=0}^{m} c_n L_n(x)$, where $c_n = \int_0^{\infty} e^{-x} x^m L_n(x) dx = (-1)^n \frac{m! m!}{n!(m-n)!}$.

4.26 The surface $\varphi = \pi/2$, corresponding to the xy-plane.

4.28 Use Eq. (4.42) and the boundary condition to obtain

$$a_n = \frac{2n+1}{2R^n} \int_0^{\pi/2} 10 P_n(\cos \varphi) \sin \varphi \, d\varphi$$

$$= \frac{5(2n+1)}{R^n} \int_0^1 P_n(x) dx = \frac{5}{R^n} [P_{n-1}(0) - P_{n+1}(0)], \quad n \in \mathbb{N},$$

where the result of Exercise 4.7 is used in the last equality for $n \in \mathbb{N}$, and $a_0 = 5$. Noting that $P_m(0) = 0$ for odd m, and $P_{2m}(0) = (-1)^m (2m)!/2^{2m} (m!)^2$, we obtain $a_1 = 15/2$, $a_3 = -35/8$, $a_5 = 55/16$, $\cdots$.

$$u(r, \varphi) = 5 \left[1 + \frac{3}{2} \frac{r}{R} P_1(\cos \varphi) - \frac{7}{8} \left(\frac{r}{R} \right)^3 P_3(\cos \varphi) + \frac{11}{16} P_5(\cos \varphi) \mp \cdots \right].$$

$u(R, \varphi) - 5$ is an odd function of φ, hence the summation starting with $n = 1$ is over odd values of n.

4.30 f may be extended as an even function of φ from $[0, \pi/2]$ to $[0, \pi]$. By symmetry the solution is even about $\varphi = \pi/2$, hence the summation is over even orders of the Legendre polynomials.

Chapter 5

5.1 For all $n \in \mathbb{N}$, the integral $I_n(x) = \int_0^n e^{-t} t^{x-1} dt$ is a continuous function of $x \in [a, b]$, where $0 < a < b < \infty$. Because

$$0 \le \int_n^\infty e^{-t} t^{x-1} dt \le \int_n^\infty e^{-t} t^{b-1} dt \to 0 \text{ as } n \to \infty,$$

it follows that I_n converges uniformly to $\Gamma(x)$. Therefore $\Gamma(x)$ is continuous on $[a, b]$ for any $0 < a < b < \infty$, and hence on $(0, \infty)$. Similarly, its derivatives are all continuous on $(0, \infty)$.

5.3 $\Gamma\left(n + \frac{1}{2}\right) = \left(n - \frac{1}{2}\right) \cdots \left(\frac{1}{2}\right) \Gamma\left(\frac{1}{2}\right) = \dfrac{(2n)!}{n! 2^{2n}} \Gamma\left(\frac{1}{2}\right)$. From Exercise 5.2 we know that $\Gamma\left(\frac{1}{2}\right) = \sqrt{\pi}$.

5.5 Replace t by $u/(1 + u)$ with appropriate limits on u.

5.6 $\operatorname{erf}(-x) = -\operatorname{erf}(x)$. e^{-t^2} is analytic in $\mathbb{R}$.

5.7 Apply the ratio test.

5.9 Differentiate Eq. (5.12) and multiply by x.

5.11 Use Eq. (5.12) to prove the identity, then substitute $\nu = -1/2$ and use Exercise 5.8.

5.13 Use $\left[x^{-\nu} J_\nu(x)\right]' = -x^{-\nu} J_{\nu+1}(x)$ and $\left[x^\nu J_\nu(x)\right]' = x^\nu J_{\nu-1}(x)$ from Example 5.2 and Exercise 5.11. Multiply the first by x^ν and the second by $x^{-\nu}$ and subtract.

5.17 Follows directly from the series representation of Y_n.

5.21 Under the transformation $x \to ix$ Bessel's equation takes the form $x^2 y'' + xy' - (x^2 + \nu^2) y = 0$. Use the relation $J_\nu(ix) = i^\nu I_\nu$.

5.24 The definition of I_ν, as given in Exercise 5.21, extends to negative values of ν. Equation (5.18) is invariant under a change of sign of ν, hence it is satisfied by both I_ν and $I_{-\nu}$.

5.26 Follows from the bounds on the sine and cosine functions.

5.28 Applying Parseval's relation to Eqs. (5.22) and (5.23), we obtain

$$\int_{-\pi}^\pi \cos^2(x \sin \theta) d\theta = 2\pi J_0^2(x) + 4\pi \sum_{m=1}^\infty J_{2m}^2(x)$$

$$\int_{-\pi}^\pi \sin^2(x \sin \theta) d\theta = 4\pi \sum_{m=1}^\infty J_{2m-1}^2(x).$$

By adding these two equations we arrive at the desired identity.

5.30 Apply Lemma 3.7 to Eqs. (5.24) and (5.25).

5.31 $f(x) = \sum_{k=1}^{\infty} c_k J_0(\mu_k x)$, $c_k = \dfrac{2}{b^2 J_1^2(\mu_k b)} \int_0^b f(x) J_0(\mu_k x) x\, dx$.

(a) $\langle 1, J_0(\mu_k x)\rangle_x = \int_0^b J_0(\mu_k x) x\, dx = \dfrac{b}{\mu_k} J_1(\mu_k b)$, $\|J_0(\mu_k x)\|_x^2 =$

$\dfrac{b^2}{2} J_1^2(\mu_k b)$. Therefore $1 = \dfrac{2}{b} \sum_{k=1}^{\infty} \dfrac{1}{\mu_k J_1(\mu_k b)} J_0(\mu_k x)$.

(e) $c_k = \dfrac{1}{b} \dfrac{J_1(\mu_k b/2)}{\mu_k J_1^2(\mu_k b)}$.

5.33 From Exercises 5.13 and 5.14(a) we have $\langle x, J_1(\mu_k x)\rangle_x = \int_0^1 J_1(\mu_k x) x^2 dx = -J_0(\mu_k)/\mu_k = J_2(\mu_k)/\mu_k$, and, from Eq. (5.34), $\|J_1(\mu_k x)\|_x^2 = \frac{1}{2} J_2^2(\mu_k)$. Therefore

$$x = 2 \sum_{k=1}^{\infty} \dfrac{1}{\mu_k J_2(\mu_k)} J_1(\mu_k x), \quad 0 \le x \le 1.$$

5.35 Using the results of Exercises 5.13 and 5.14(a),

$$\langle f, J_1(\mu_k x)\rangle_x = \int_0^1 x^2 J_1(\mu_k x) dx = \dfrac{1}{\mu_k} J_2(\mu_k),$$

$$f(x) = \sum_{k=1}^{\infty} \dfrac{J_2(\mu_k)}{2\mu_k J_1^2(2\mu_k)} J_1(\mu_k x), \quad 0 \le x \le 2.$$

This representation is not pointwise. At $x = 1$, $f(1) = 1$ and the right-hand side is $\frac{1}{2}[f(1^+) + f(1^-)] = \frac{1}{2}$.

5.37 Assuming $u(r, t) = v(r)w(t)$ leads to $\dfrac{w'}{kw} = \dfrac{1}{v}\left(v'' + \dfrac{1}{r}v'\right) = -\mu^2$. Solve these two equations and apply the boundary condition to obtain the desired representation for u.

5.39 Use separation of variables to conclude that

$$u(r, t) = \sum_{k=1}^{\infty} J_0(\mu_k r)(a_k \cos \mu_k ct + b_k \sin \mu_k ct),$$

$$a_k = \dfrac{2}{R^2 J_1^2(\mu_k R)} \int_0^R f(r) J_0(\mu_k r) r\, dr,$$

$$b_k = \dfrac{2}{c\mu_k R^2 J_1^2(\mu_k R)} \int_0^R g(r) J_0(\mu_k r) r\, dr.$$

Chapter 6

6.1 (a) $\hat{f}_1(\xi) = \frac{2}{\xi^2}(1 - \cos \xi)$. (b) $\hat{f}_2(\xi) = 2\dfrac{\sin \xi}{\xi}$.

6.2 (a) $\mathcal{L}^1 \cap \mathcal{L}^2$. (b) $\mathcal{L}^2$.

6.3 $2\left[1 + (\xi - a)^2\right]^{-1}$.

6.5 $F(\xi) = \pi/2$ when $\xi > 0$, $F(\xi) = -\pi/2$ when $\xi < 0$, and $F(0) = 0$, so the indefinite integral converges to an odd function with a jump discontinuity at $\xi = 0$ of magnitude π.

6.8 $\hat{f}(\xi) = 2\dfrac{\xi \sin \pi \xi}{1 - \xi^2}$. Not in $\mathcal{L}^1(\mathbb{R})$.

6.9 Express the integral over (a, b) as a sum of integrals over the sub-intervals $(a, x_1), \cdots, (x_n, b)$. Since both f and g are smooth over each sub-interval, the formula for integration by parts applies seperately to each integral in the sum.

6.10

(a) f is even, therefore $B(\xi) = 0$, $A(\xi) = 2\displaystyle\int_0^\pi \sin x \cos \xi x \, dx = 2\dfrac{1 + \cos \pi \xi}{1 - \xi^2}$,

and $f(x) = \dfrac{2}{\pi}\displaystyle\int_0^\infty \dfrac{1 + \cos \pi \xi}{1 - \xi^2} \cos x\xi \, d\xi$.

(b) $A(\xi) = \dfrac{\sin \xi}{\xi}$, $B(\xi) = \dfrac{1 - \cos \xi}{\xi}$, hence

$$f(x) = \frac{1}{\pi}\int_0^\infty \left(\frac{\sin \xi}{\xi} \cos x\xi + \frac{1 - \cos \xi}{\xi} \sin x\xi \right) d\xi.$$

6.12 Let f be the odd extension of $e^{-\alpha x}$ from $(0, \infty)$ to $\mathbb{R}$ and $\alpha > 0$, then its Fourier sine transform is

$$B(\xi) = \frac{2\xi}{\xi^2 + \alpha^2} \quad \text{and hence} \quad f(x) = \frac{2}{\pi}\int_0^\infty \frac{\xi}{\xi^2 + \alpha^2} \sin x\xi \, d\xi = e^{-x}$$

for all $x > 0$. At $x = 0$, $f(x)$ is discontinuous and the integral equals $\frac{1}{2}\left[f(0^+) + f(0^-)\right] = 0$.

6.15 Extend

$$f(x) = \begin{cases} 1, & 0 < x < \pi \\ 0, & x > \pi \end{cases}$$

as an odd function to $\mathbb{R}$ and show that its sine transform is $B(\xi) = 2(1 - \cos \pi \xi)/\xi$.

6.17 Show that the cosine transform of f is

$$A(\xi) = 2\frac{1 - \cos \xi}{\xi^2} = \frac{\sin^2(\xi/2)}{(\xi/2)^2}.$$

Express $f(x)$ as a cosine integral and evaluate the result at $x = 0$, which is a point of continuity of f.

6.19 Equations (6.21) and (6.31) imply that $\left\| \hat{f} \right\|^2 = \|A\|^2 + \|B\|^2 = 2\pi \|f\|^2$.

6.20 $\psi_n(x)$ decays exponentially as $|x| \to \infty$, so it belongs to $\mathcal{L}^1(\mathbb{R})$ and $\hat{\psi}$ therefore exists. From Example 6.17 we have $\hat{\psi}_0(\xi) = \sqrt{2\pi}\,\psi_0(\xi)$. Assuming $\hat{\psi}_n(\xi) = (-i)^n \sqrt{2\pi}\,\psi_n(\xi)$, we have

$$\hat{\psi}_{n+1}(\xi) = \mathcal{F}\left(e^{-x^2/2} H_{n+1}(x)\right)(\xi)$$

$$= \mathcal{F}\left[e^{-x^2/2}(2x H_n(x) - H_n'(x))\right](\xi)$$

$$= \mathcal{F}\left[x\psi_n(x) - \psi_n'(x)\right](\xi)$$

$$= i\hat{\psi}_n'(\xi) - i\xi\hat{\psi}_n(\xi)$$

$$= (-i)^{n+1}\sqrt{2\pi}\,[-\psi_n'(\xi) + \xi\psi_n(\xi)]$$

$$= (-i)^{n+1}\sqrt{2\pi}\,\psi_{n+1}(\xi),$$

where we used the identity $H_{n+1}(x) = 2x H_n(x) - H_n'(x)$ and Theorem 6.15. Thus, by induction, $\hat{\psi}_n(\xi) = (-i)^n \sqrt{2\pi}\,\psi_n(\xi)$ is true for all $n \in \mathbb{N}_0$.

6.22 Define the integral $I(z) = \int_0^\infty e^{-b\xi^2} \cos z\xi \, d\xi$ and show that it satisfies the differential equation $I'(z) = -zI(z)/2b$. Solve for $I(z)$ and evaluate at 0.

6.23 The general solution of the heat equation is

$$u(x, t) = \frac{1}{\pi} \int_0^\infty [A(\lambda) \cos \lambda x + B(\lambda) \sin \lambda x]\, e^{-k\lambda^2 t}\, d\lambda.$$

The boundary condition at $x = 0$ implies $A(\lambda) = 0$. By extending $f(x)$ as an odd function from $(0, \infty)$ to $(-\infty, \infty)$ we see that $B(\lambda)$ is the sine transform of f, and the same procedure followed in Example 6.18 leads to the desired result.

6.25 Solve the transformed wave equation $\hat{u}_{tt}(\xi, t) = -c^2\xi^2\hat{u}(\xi, t)$, under the given initial conditions $\hat{u}(\xi, 0) = \hat{f}(\xi)$, $\hat{u}_t(\xi, 0) = 0$, then invert $\hat{u}(\xi, t)$.

Chapter 7

7.1

(a) $\dfrac{2a^2}{s^3} + \dfrac{2ab}{s^2} + \dfrac{b^2}{s}$.

(d) $\dfrac{1}{s^2 + 4}$.

(f) $\sqrt{\pi/s}$.

7.2

(a) $e^x \cos 2x$.

(b) $2\cosh 3x - \dfrac{5}{3}\sinh 3x$.

(d) $2\sqrt{x/\pi}$.

7.4

(e) $s^{-1}\left(1 - e^{-s}\right) - (s + 1)^{-1}\left[1 - e^{-(s+1)}\right]$.

7.5 $f(x) = x[H(x) - H(x - 1)] + e^{1-x}H(x - 1)$.

$$\mathcal{L}(f)(\xi) = \frac{1}{s^2}\left(1 - e^{-s}\right) - \frac{1}{s}e^{-s} + \frac{1}{s+1}e^{-s}.$$

7.6

(c) $H(x - 3) + H(x - 1)$.

(d) $H(x - 3)\,e^{x-3} + H(x - 1)\,e^{x-1}$.

7.7 For each point of jump discontinuity x_i the formula should be corrected by adding $\left[f\left(x_i^-\right) - f\left(x_i^+\right)\right]e^{-sx_i}$ to the right-hand side of the equality.

7.8

(a) $y = e^{-2x}\left(\cos x + 3\sin 2x\right)$.

(e) $y = \left(e^{2x} - e^x\right)H(x) + \left(\frac{1}{2}e^{2(x-1)} - e^{x-1} + \frac{1}{2}\right)H(x - 1)$.

7.9

(a) $\mathcal{L}^{-1}\left[\dfrac{s}{(s^2 + 9)^2}\right] = \dfrac{1}{6}\mathcal{L}^{-1}\left[-\dfrac{d}{ds}\left(\dfrac{3}{s^2 + 9}\right)\right] = \dfrac{1}{6}x\sin 3x$.

(b) $\dfrac{1}{x}\left(e^x - 1\right)$.

(c) $\dfrac{1}{x}\left(e^{-bx} - e^{-ax}\right)$.

(d) Use Eq. (7.13) to obtain $\mathcal{L}^{-1}\left[(\operatorname{arccot}(s + 1)\right] = e^{-x}\mathcal{L}^{-1}(\operatorname{arccot}s)$, then show that $\mathcal{L}^{-1}(\operatorname{arccot}s) = \sin x/x$.

7.10 $s\mathcal{L}(\text{Si})(s) = \mathcal{L}\left(\dfrac{d}{dx}\text{Si}(x)\right)(s) = \mathcal{L}\left(\dfrac{\sin x}{x}\right)(s) = \arctan\left(\dfrac{1}{s}\right)$.

7.12 For any $f \in \mathcal{E}$ there are positive constants M and N such that $|f(x)| \le Me^{\alpha x}$ on (N, ∞) for some $\alpha \ge 0$. Therefore, setting $\operatorname{Re}s = \sigma$,

$$|\mathcal{L}(f)(s)| \le \int_0^\infty |f(x)|\,e^{-\sigma x}dx$$

$$\le \int_0^N |f(x)|\,e^{-\sigma x}dx + M\int_N^\infty e^{-(\sigma - \alpha)x}dx$$

$$\le \int_0^N |f(x)|\,e^{-\sigma x}dx + \frac{M}{\sigma - \alpha}e^{-(\sigma - \alpha)N}, \quad \sigma > \alpha.$$

Since f is locally integrable, the right-hand side tends to 0 as $\sigma \to \infty$.

7.13 The transformed equation is $6Y(s)/s^4 = F(s)$, from which $Y(s) = s^4 F(s)/6$. Using Theorem 7.6 and its corollary we conclude that

$$y(x) = \frac{1}{6} f^{(4)}(x) + \frac{1}{6} \mathcal{L}^{-1}[f(0^+)s^3 + f'(0^+)s^2 + f''(0^+)s + f'''(0^+)].$$

The integral expression for $f(x)$ implies that $f^{(n)}(0^+) = 0$ for $n = 0, 1, 2, 3$, hence $y(x) = f^{(4)}(x)/6$.

7.14 $f * g(x) = \displaystyle\int_0^x f(x-t)\, g(t)\, dt = \int_0^x e^{x-t} \frac{1}{\sqrt{\pi t}} dt = e^x \frac{1}{\sqrt{\pi}} \int_0^x e^{-t} \frac{1}{\sqrt{t}} dt.$

Under the change of variable $t = p^2$ we obtain $f * g(x) = e^x \mathrm{erf}\sqrt{x}$. Now $\mathcal{L}\left(e^x \mathrm{erf}\sqrt{x}\right) = \mathcal{L}(e^x)\, \mathcal{L}(g) = \dfrac{1}{s-1}\dfrac{1}{\sqrt{s}}$, from which $\mathcal{L}\left(\mathrm{erf}\sqrt{x}\right) = \dfrac{1}{s\sqrt{s+1}}$.

7.17 The bounded solution of the transformed wave equation $U_{xx} = s^2 U/c^2$ is $U(s) = U(0, s)\, e^{-sx/c}$, therefore

$$u(x, t) = \mathcal{L}^{-1}\left[e^{-sx/c} \mathcal{L}\left(\cos^2 t\right)\right] = H(t - x/c) \cos^2(t - x/c).$$

7.19 $u(x, t) = \mathcal{L}^{-1}\left[e^{-(s+1)x/c} \mathcal{L}(\sin t)\right] = e^{-x/c} H(t - x/c) \sin(t - x/c).$

7.21 $F(s) = e^{-a\sqrt{s}}/\sqrt{s}$ is analytic in the complex plane cut along the negative axis $(-\infty, 0]$. Using Cauchy's theorem, the integral along the vertical line $(\beta - i\infty, \beta + i\infty)$ can be reduced to two integrals, one along the bottom edge of the cut from left to right, and the other along the top edge from right to left. This yields

$$\begin{aligned}
\mathcal{L}^{-1}(F)(x) &= \frac{1}{2\pi i} \int_{\beta - i\infty}^{\beta + i\infty} \frac{e^{-a\sqrt{s}}}{\sqrt{s}} e^{sx}\, ds \\
&= \frac{1}{2\pi} \int_0^\infty \frac{e^{ia\sqrt{s}} + e^{-ia\sqrt{s}}}{\sqrt{s}} ds \\
&= \frac{1}{\pi} \int_0^\infty \frac{\cos a\sqrt{s}}{\sqrt{s}} e^{-sx}\, ds = \frac{2}{\pi} \int_0^\infty e^{-xt^2} \cos at\, dt.
\end{aligned}$$

Noting that the last integral is the Fourier transform of e^{-xt^2}, and using the result of Example 6.17, we obtain the desired expression for $\mathcal{L}^{-1}(F)(x)$. To evaluate $\mathcal{L}^{-1}\left(e^{-a\sqrt{s}}\right)$ we first integrate over (s, ∞) to eliminate $\sqrt{s}$ from the denominator of F,

$$\int_s^\infty \frac{e^{-a\sqrt{z}}}{\sqrt{z}} dz = 2 \int_{\sqrt{s}}^\infty e^{-at}\, dt = \frac{2}{a} e^{-\sqrt{s}}, \quad a > 0,$$

then apply Theorem 7.6 to obtain

$$\mathcal{L}^{-1}\left(e^{-a\sqrt{s}}\right)(x) = \frac{a}{2}\mathcal{L}^{-1}\left[\int_s^\infty \frac{e^{-a\sqrt{z}}}{\sqrt{z}}dz\right] = \frac{a}{2x}\mathcal{L}^{-1}\left(\frac{e^{-a\sqrt{s}}}{\sqrt{s}}\right)(x)$$

$$= \frac{a}{2\sqrt{\pi x^3}}e^{-a^2/4x}.$$

Appendix

A.1 For any $x \in (0, 1]$, $1 - x^2 < 1$ hence $\lim_{n\to\infty} nx\left(1 - x^2\right)^n = 0$. Therefore the integral of the limit $\int_0^1 \lim_{n\to\infty} nx\left(1 - x^2\right)^n dx$ is 0. On the other hand, $\int_0^1 nx\left(1 - x^2\right)^n dx = -\frac{n}{2}\int_0^1 \left(1 - x^2\right)^n (-2x)\,dx = \frac{n}{2(n+1)} \to \frac{1}{2}$ as $n \to \infty$.

A.2 If $c = 0$ then $cf = 0$ is measurable. If $cf \neq 0$, then for all $\alpha \in \mathbb{R}$

$$\left\{x \in \overline{\mathbb{R}} : cf(x) > \alpha\right\} = \begin{cases} \left\{x \in \overline{\mathbb{R}} : f(x) > \alpha/c\right\} & \text{if } c > 0, \\ \left\{x \in \overline{\mathbb{R}} : f(x) < \alpha/c\right\} & \text{if } c < 0, \end{cases}$$

and both sets are measurable, hence so is cf. The converse is false, since cf is measurable if $c = 0$ even if f is not measurable.

A.3 If $\alpha > 0$, $\left\{x \in \overline{\mathbb{R}} : |f(x)| > \alpha\right\} = \left\{x \in \overline{\mathbb{R}} : f(x) > \alpha\right\} \cup \left\{x \in \overline{\mathbb{R}} : f(x) < -\alpha\right\}$ is measurable, and if $\alpha < 0$ then $\left\{x \in \overline{\mathbb{R}} : |f(x)| > \alpha\right\} = \overline{\mathbb{R}}$ is also measurable. The converse is false, for if E is a non-measurable proper subset $\mathbb{R}$, then the function $f(x)$ defined as 1 on E and -1 on $\mathbb{R}\backslash E$ is non-measurable, though $|f|$ is measurable.

A.4 By the completeness of $\mathbb{Q}$ in $\mathbb{R}$, for any irrational number x in $[0, 1]$ there is a sequence (x_n) in $\mathbb{Q} \cap [0, 1]$ such that $x = \lim x_n$. But $f(x) = 0$, whereas $f(x_n) = 1$ for all n, so f is not continuous at any point in the set $\mathbb{Q}^c \cap [0, 1]$, whose measure is $m([0, 1]) - m(\mathbb{Q} \cap [0, 1]) = 1$.

A.5 Define $f_n(x) = \chi_{[0,n]}(x)(1 - x/n)^n e^{-x}$, so that $\int_0^n (1 - x/n)^n e^{-x}dx = \int_0^\infty f_n(x)\,dx$. Now $|f_n(x)| \leq e^{-x} \in \mathcal{L}^1(0, \infty)$ for all n. By the dominated convergence theorem and the convergence of $(1 - x/n)^n$ to e^{-x}, $\lim_{n\to\infty} \int_0^n (1 - x/n)^n e^{-x}dx = \int_0^\infty \lim f_n(x)\,dx = \int_0^\infty e^{-2x}dx = 1/2$.

References

1. Al-Gwaiz, M.A., Elsanousi, S.A.: Elements of Real Analysis. Chapman and Hall/CRC, Boca Raton (2006)
2. Bartle, R.G., Sherbert, D.R.: Introduction to Real Analysis. John Wiley, New York (1982)
3. Birkhoff, G., Rota, G.-C.: Ordinary Differential Equations, 2nd edn. John Wiley, New York (1969)
4. Carslaw, H.S.: Introduction to the Theory of Fourier's Series and Integrals, 3rd edn. Dover, New York (1930)
5. Churchill, R.V., Brown, J.W.: Fourier Series and Boundary Value Problems, 6th edn. McGraw-Hill International, New York (2001)
6. Coddington, E.A., Levinson, N.: Theory of Ordinary Differential Equations. McGraw-Hill, New York (1955)
7. Courant, R., Hilbert, D.: Methods of Mathematical Physics, vols. I and II. Interscience Publishers, New York (1953/1963)
8. Courant, R., John, F.: Introduction to Calculus and Analysis, vol. II. John Wiley, Hoboken (1974)
9. Folland, G.B.: Fourier Analysis and Its Applications. Wadsworth, Belmont (1992)
10. González-Velasco, E.A.: Fourier Analysis and Boundary Value Problems. Academic Press, San Diego (1995).
11. Halmos, P.R.: Finite-Dimensional Vector Spaces, 2nd edn. Van Nostrand, Princeton (1958)
12. Ince, E.L.: Ordinary Differential Equations. Dover, New York (1956)
13. John, F.: Partial Differential Equations, 4th edn. Springer, New York (1982)
14. Royden, H.L.: Real Analysis, 3rd edn. Macmillan, New York (1988)
15. Rudin, W.: Principles of Mathematical Analysis. McGraw-Hill, New York (1964)
16. Titchmarch, E.C.: Eigenfunction Expansions Associated with Second Order Differential Equations, 2nd edn. Clarendon Press, Oxford (1962)
17. Tolstov, G.P.: Fourier Series. Dover, New York (1962)
18. Walter, W.: Ordinary Differential Equations. Springer, New York (1998)
19. Watson, G.N.: A Treatise on the Theory of Bessel Functions, 2nd edn. Cambridge University Press, Cambridge (1944)
20. Zettl, A.: Sturm-Liouville Theory, vol. 121. American Mathematical Society Publication, Providence (2005)

© The Author(s), under exclusive license to Springer-Verlag London Ltd., part of Springer Nature 2026

M. A. Al-Gwaiz, *Sturm-Liouville Theory and its Applications*, Springer Undergraduate Mathematics Series, https://doi.org/10.1007/978-1-4471-7610-7

Index

MIX
Papier aus verantwortungsvollen Quellen
Paper from responsible sources
FSC® C105338

If you have any concerns about our products,
you can contact us on
ProductSafety@springernature.com

In case Publisher is established outside the EU,
the EU authorized representative is:
Springer Nature Customer Service Center GmbH
Europaplatz 3, 69115 Heidelberg, Germany

Printed by Libri Plureos GmbH
in Hamburg, Germany

MIX
Papier aus verantwortungsvollen Quellen
Paper from responsible sources
FSC® C105338
FSC
www.fsc.org